U0165700

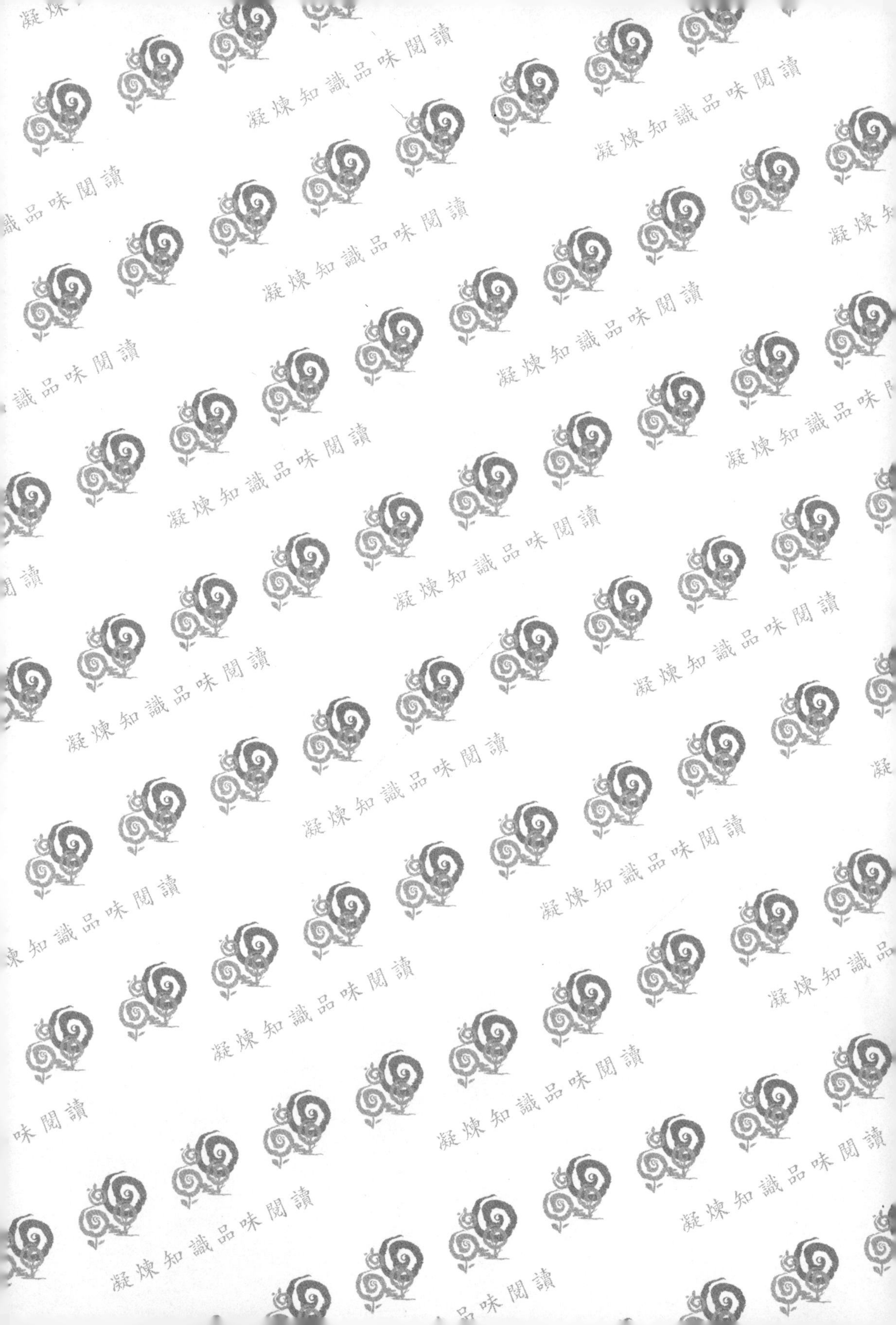

凝煉知識品味閱讀

應用外語21

商用英文寫作

ENGLISH

五南圖書出版公司 印行

朱海成・著

PREFACE

作者簡介

朱海成教授

●作者背景

★現職

國立臺中教育大學／管理學院／國際企業學系／專任教授

國立臺中教育大學國際長／研發長／兩岸交流處處長

美國紐約理工學院（New York Institute of Technology）在臺商用/科技英語認證教授

鄭州大學西亞斯國際學院（SIAS International University）榮譽講座教授

★學歷

美國紐約州立大學（SUNY）at Binghamton系統科學與工業工程博士

★經歷

1. 美國紐約Cheyenne Software Inc. 資深軟體工程師（New York, USA）（1996）
2. 榮獲公費赴美國哈佛大學商學院進修(Harvard University - Business School / PCMPCL V Graduated)(2007), MA, USA.

3. 加拿大卑詩省UNBC管理學院暑期教師（2008）
4. 榮獲僑委會公費赴美國西雅圖、加拿大溫哥華、多倫多巡迴輔導台商資訊管理／電子商務企業實務運作（2002）
5. 榮獲外交部邀請至外交部五樓大禮堂專題演講－外貿在外交之應用（1999）
6. 臺灣臺中軟體園區發展產學訓聯盟發起人－國立臺中教育大學代表（2014）
7. 拜訪蒙古國（烏蘭巴托市）、日本（京都、大阪、東京市）、韓國（首爾市、光州市）、印尼（雅加達市、泗水市）、馬來西亞（吉隆坡市）、寮國（永珍市）、柬埔寨（金邊市）、泰國（曼谷、孔敬）、越南（河內、胡志明市）、丹麥（哥本哈根）、俄羅斯（莫斯科、聖彼得堡）、英國（倫敦、OXFORD校長室）、美國（UCLA校長室、BYU校長室、NYIT校長室、NJIT校長室、Kent State University校長室、Ashland校長室）、加拿大（Royal Roads University校長室、Thompson River University校長室）、中國（哈爾濱市、長春市、大連市、北京市、上海市、長熟市、杭州市、鄭州市、新鎮市、福州市、廈門市、泉州市、漳州市、武漢市、長沙市、襄陽市、廣州市、湛江市、西安市，香港、澳門等），與當地教育、商務組織交流密切。
8. 曾任教於逢甲國貿、東海國貿、東海EMBA連續8年。
9. 新竹台積電、旺宏電子、台中世貿中心系列授課。

PREFACE

作者序

【商用英文寫作】的內容主要是以台灣國際企業相關商務主題，而撰寫之商用英文書信寫作範本，包含與國外公司、海外客戶等日常業務聯繫書信，讀者可在短時間內吸收並加以活用。信函的種類有：歡迎信、邀請函、自我介紹、錄用函、拒絕函、傳真信函、推薦函、申請函、工作面試函、重新預約函、樣品需求函、搬遷函、病假函、徵聘函、簽署備忘錄函、週年紀念函、嬰兒誕生函、致歉函、弔唁函、告別函、談判函、衝突解決信函、意外事件／事故信函、詢問報價信函、下訂單函、物料需求規劃信函、製造資源規劃信函、調整報價函、追蹤函、飛機票預約函、參展邀請函、旅館預約函、租車函、交通延遲函、開發新客戶函、留住舊顧客函、顧客滿意度信函、尋求顧客忠誠度信函、處理顧客抱怨信函…等等。

本書特地自常用英文文法切入，讓讀者熟悉相關之常用文法規則，避免不成熟的商務英文寫作，同時在每個重要單元結束時，還提供實戰寫作技巧，藉以快速提升讀者之商務英文寫作能力。

作者曾以相同教材與方法在新竹台積電、旺宏電子、台中世貿中心系列授課，廣受好評。現為美國紐約理工學院（New York Institute of Technology）在臺商用／科技英語認證教授。

國立臺中教育大學管理學院／國際企業學系

朱海成 教授 敬上

ayura66@gmail.com

CONTENTS

目錄

Module 1
Essential Writing Structure Review（基礎寫作架構復習）

Module 2
Business Writing Essentials（基本商業書信）

目錄

Module 3
Special Occasions（特殊場合）

Module 4
Quotations, Outsourcing, and other e-Commerce Occasions（報價、委外、其它電子商務場合）

目錄

Module 6
Customer Relationship Management（顧客關係管理信函）

目錄

Module 7

English Proverbs（英文諺語）

Module 8

Eyes Catching Business Slogans（吸睛的商業標語）

Module 01

Essential Writing Structure Review
基礎寫作架構復習

Chapter ❶ 關係代名詞（Relative Pronoun）

Chapter ❷ 動名詞（Gerund）

Chapter ❸ 不定詞（Infinitive）

Chapter ❹ 時態（Tense）

Chapter ❺ 比較級與最高級（Comparative & Superlative）

Chapter ❻ 代名詞（Pronoun）

Chapter ❼ 連接詞（Conjunction）

Module 01

關係代名詞（Relative Pronoun）

▶ 關係代名詞有who、whom、which，均用以引導形容詞子句，修飾前面之名詞。

1. 關係代名詞為【主格】時之用法：

Example:

他是偷你錢的人。

He is the man who stole your money.

2. 關係代名詞為【受格】時之用法：

Example:

我能相信你那麼恨的那個人嗎？

Can I trust a man whom you hate so much?

▶ 關係代名詞的使用時機

1. 關係代名詞之前要有名詞（Nouns）。
2. 關係代名詞所引導的子句中，本身要做主詞（主格）或受詞（受格）。否則關係代名詞之前要有介係詞，而該介係詞亦可接至句尾。

Example:

他是我能信任的人。

He is a man whom I can trust.

Example：

他是我樂於替他工作的人。

He is a man whom I enjoy working for.

Example：

他是我樂於共事的人。

He is a man with whom I enjoy working.

介係詞

或

He is a man whom I enjoy working with.

介係詞放到句尾

▶ 關係代名詞【非限定用法】與關係代名詞【限定用法】之使用時機

1.【非限定用法】修飾前面之主詞：

專有名詞如John、Peter、U.S.A、my father、my mother等具唯一性的名詞，則不需加以限定。此時關係代名詞所引導之形容詞子句，之前要加逗點。

Example：

我喜歡John，他工作認真。

I like John, who works hard.

Example：

我看見我媽，我非常愛她。

I saw my mother, whom I love very much.

Example：

我見到我深愛的那位媽媽。

I saw my mother whom I love very much. → 此說法與常理不符

2. 【限定用法】修飾前面之主詞：

Example：

我知道那家以中式料理聞名的餐館。

I know the restaurant which is very famous for its Chinese cuisine.

▲ 在很多場合，that可用來取代who、whom或which，但有2個使用原則：

A. 其前不可有介係詞。

B. 其前不可有逗點。

Example：

我喜歡John，他工作認眞。

I like John, that works hard. (×)

I like John, who works hard. (○)

Example：

他是我樂於替他工作的人。

He is the man for that I enjoy working. (×)

He is the man for whom I enjoy working. (○)

▶ 若that之前有插入語，則不受逗點之限制。

Example：

He is a good boy, as far as I know, that (whom) you can trust.

▶ which亦可代替前面整個子句。

Example：

He is a naughty boy, which everyone knows.

▶ 做為受詞的whom或which（含that）在限定用法修飾的句構中，可以省略。

Example：

他是我可以信賴的人。

He is a man whom (that) I can trust. = He is a man I can trust.

Example：

這是我昨天遺失的書。

This is the book which (that) I lost yesterday.

= This is the book I lost yesterday.

Example：

他是我樂於替他工作的人。

He is the man whom (that) I enjoy working for.

= He is the man I enjoy working for.

▶ 句子與句子間一定要有連接詞（Conjunction）或標點符號（Punctuation）。

Example：

He is a good boy, I like him. (×) →無連接詞

He is a good boy. Because he studies hard. (×) →Because為連接詞，但其前是句點。

He is a good boy and I like him. (○)

▶ 如何連接兩個句子

1. 分號；

   Example：

   他是個好男孩，我喜歡他。

   He is a good boy; I like him.

2. 破折號（Hyphen）－

   Example：

   他是個好男孩，他認真學習。

   He is a good boy － he studies very hard.

3. 連接詞　例如：so或because

   Example：

   他是個好男孩，所以我喜歡他。

   He is a good boy, so I like him.

   Example：

   他是個好男孩，因爲他認眞學習。

   He is a good boy, because he studies hard.

4. 關係詞　例如：whom

Example：

他是個好男孩，我非常喜歡他。

He is a good boy, whom I like very much.

▶ 若兩個句子之間無連接詞，第一個子句改成分詞構句。

步驟：

1. 去掉相同主詞。
2. V→Ving（任何動詞改成ing形式）。
3. being or having been 則可省略。

Example：

他是個好男孩，他認眞學習。

He is a nice boy, he studies hard. (×)

(Being) A nice boy, he studies hard. (○)

Example：

他有重要的事做，他無法和我們一起去露營。

He had something important to do, he couldn't go camping with us. (×)

Having something important to do, he couldn't go camping with us. (○)

Example：

我到了臺北，我注意到下雨了。

I arrived in Taipei, I noticed that it was rainy. (×)

Arriving in Taipei, I noticed that it was rainy. (○)

Example：

她穿著白色衣服，她參加了派對。

She was dressed in white, she attended the party. (×)

(Being) Dressed in white, she attended the party. (○)

▶ 使用分詞構句之否定句時，否定副詞not或never應放在分詞之前。

Example：

他不喜歡學習，他逃家。

He was not fond of learning, he ran away from home. (×)

Not being fond of learning, he ran away from home. (○)

▶ 由when、while、once、if、unless、though等副詞連接詞所引導之副詞子句，若其主詞與主要子句相同時，可保留該副詞連接詞，其餘部分簡化成分詞片語。

Example：

當我有空時，會與你連絡。

When I am free, I'll contact with you.

副詞連接詞所引導之副詞子句

When (being) free, I'll contact with you.

Example:

一旦我到那裡，我會與你連絡。

Once I arrive there, I'll keep contacting with you.

副詞連接詞所引導之副詞子句

Once arriving there, I'll keep contact with you.

Example:

雖然他知道事實，他仍然沉默。

Though he knew the truth, he remained silent.

副詞連接詞所引導之副詞子句

Though knowing the truth, he remained silent.

獨立分詞構句：

▶ **使用時機：在分詞構句中，若前後主詞不同，則第一個子句之主詞予以保留。**

Example:

放學了，所有的孩子回家。

School was over, all the children went home. (×)

前後主詞不同

School (being) over, all the children went home. (○)

第一個子句之主詞予以保留

Example:

這男孩犯了錯，他父親生氣了。

The boy had made the mistake, his father got angry.

前後主詞不同

The boy having made the mistake, his father got angry.

第一個子句之主詞予以保留

▶ Whose為關係代名詞之所有格，由his、her、their、your以及its等所有格變化而成。與關係代名詞一樣，whose引導形容詞子句，修飾前面之名詞。

Example:

這是John，他的父親是我的英文老師。

This is John, whose father is my English teacher.

由his變化而成

Example:

他正在閱讀一本書，那本書的寫法很難令人了解。

He is reading a book whose writing style is hard to be understood.

由its變化而成

Example:

居住在城市的人，希望住在鄉下。

People whose homes are in town want to live in the country.

由their變化而成

▶ 使用 whose 為關係代名詞時，請注意以下原則:

1. whose之前要有名詞。
2. whose之後的名詞，在whose所引導的形容詞子句中，當做主詞或受詞。
3. 否則，whose之前要放介係詞（preposition），而該介係詞亦可放於句尾。

Example:

我嫉妒John，他的車很拉風。

I envy John, whose car is fancy.

形容詞子句，當做主詞

Example:

我遇到John，我不喜歡他的父親。

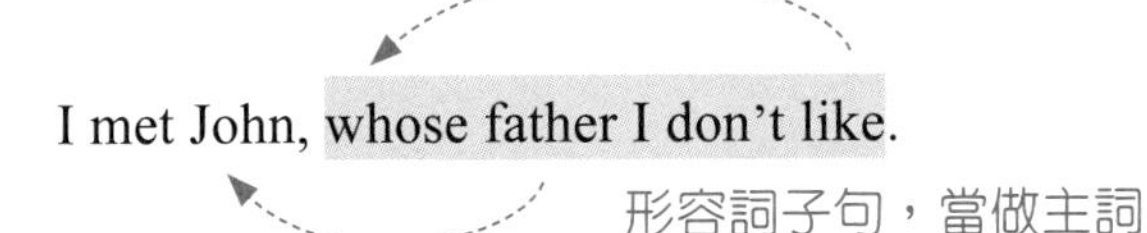

• 易犯錯誤範例

I like John, whose sister I desire to make friends. (×)

正確解析

- whose之前有名詞，滿足上述條件A。
- 但whose sister所引導之形容詞子句，無法做主詞（因已有主詞John），也無法做受詞（因已有受詞friends）。
- 故其前要有介係詞（此情況使用with），因片語：make friends with（與某人交朋友）。

正解：

I like John, whose sister I desire to make friends with. (○)

介係詞放於句尾

或

I like John, with whose sister I desire to make friends. (○)

介係詞放於whose之前

▶ What可做複合關係代名詞，在句中，整個what引導之子句，可做主詞、受詞、或be動詞補語（Complement）。

Example：

他所說的或許正確。

What he said may be true.

what引導之子句做主詞

Example：

我相信他所說的。

I believe what he said.

what引導之子句做受詞

Example：

我對他所說的有興趣。

I am interested in what he said.

what引導之子句做受詞

Example：

這是他所說的。

This is what he said.

what引導之子句做be動詞補語

▶ 使用複合關係代名詞What時，有以下使用原則：

1. what之前不可有名詞。
2. 將what詮釋爲the thing which。
3. 因爲which爲關係代名詞，在所引導之形容詞子句中，可當做主詞或受詞。因此，what所引導之形容詞子句，也可當做主詞或受詞。

• 易犯錯誤範例

Example：

這是他所做完的事。

This is the thing what he has done. (×)

➧ what之前不可有名詞。

正 解：

This is what he has done. (○)

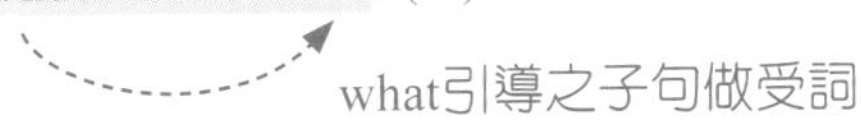

→ what在此等於the thing which，做爲done之受詞。

Example：

他所說的可能正確。

What he said may be true.

→ what在此等於the thing which，做爲said之受詞。

閃耀者未必就是金子。

All that glitters is not gold.

- all that = everything that（所有……的一切東西）。
- 換言之，all相當於everything，而that爲關係代名詞。

Example：

他所說的是眞實的。

All that he said is true.

→ Everything that he said is true.

▶ 在all that所引導之形容詞子句中，如果that當主詞，該that不可省略；如果that當受詞，則可以省略。

Example：

他所說的是眞實的。

All that he said is true.

that為said之受詞，可以省略。

→ All he said is true.

Example：

該做的事，我們應該做。

We must do all that is to be done.

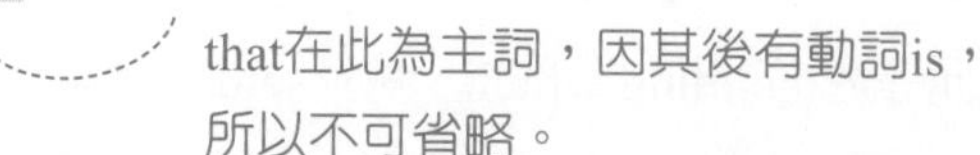

that在此為主詞，因其後有動詞is，所以不可省略。

▶ all that往往為what所取代。

Example：

他所說的是眞實的。

All that he said is true.

= What he said is true.

Example：

該做的事，我們應該做。

We must do all that is to be done.

We must do what is to be done.

▶ 在有些情況下，形容詞子句有強烈之限定意味時，其關係代名詞要用that取代who、whom或which。例如：the only、the very、the first、the last這類情況。

Example：

人類是唯一與生俱來就會說話的生物。

Man is the only creature that is gifted with speech.

Example：

他將是我最不想相處的朋友。

He is the last person that I'll get along with.

Example：

這是我見過最美麗的花朵。

This is the prettiest flower that I have ever seen.

▶ 在有些情況下，為避免與疑問句重複用詞造成贅字，可使用that取代who、whom、which。

Example:

Who is the man who is standing there?

who引導疑問詞關係代名詞

建議改寫為：

Who is the man that is standing there?

▶ 其他分詞（Participle）句型使用時機：

若有兩個動詞在同一句中，而無連接詞加以連接時，請考慮以下使用原則：

1. 若此兩個動詞，按句義來分析，若表示動作同時發生時，則將第二個動詞改成現在分詞（Ving形式）。
2. 若此兩個動作不同時發生，則第二個動詞改成不定詞（To + 原型V）。

Example:

那孩子唱著歌離開。

The children went away singing a song.

Ving形式

Example:

那隻狗在門前躺著打瞌睡。

The dog lay dozing in front of the door.

躺 lie – lay – lain – lying

說謊 lie – lied – lied – lying

Example：

John坐在角落處看報紙。

John sat in the corner reading a newspaper.

Example：

他來這裡爲的是看我。

He came here to see me.

↘ to + V形式

Example：

他努力學習爲的是通過考試。

He studied hard to pass the examination.

↘ to + V形式

句型比較

Example：

他坐下來抽菸。

He sat down to smoke.

Example：

他坐下的同時也抽菸。

He sat down smoking.

承前所述，若兩個動詞表同時進行之意味，而第二個動詞為be動詞時，仍改成現在分詞，換言之，即為being，但being通常省略。

Example：1

他站在那裡不動。

He stood there was motionless. (×)

正解：

He stood there (being) motionless. (○)

Example：2

他過世時是英雄。

He died was a hero. (×)

正解：

He died (being) a hero. (○)

Example：3

他年輕時離家。

He left home was young. (×)

正解：

He left home (being) young. (○)

▶ 若兩個動詞在一起，而第二個動詞之前有逗點時，則應將該動詞改成現在分詞。

Example：

他單手拿著一把刀，朝向May走過去。

He marched toward May, holding a knife in his hand.

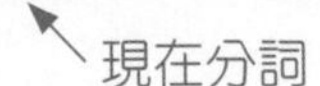

寫作練習

1. 那些女孩子們一路唱著歌來到城裡。
2. 我們坐在山頂說說唱唱。
3. 許多人在教堂前冷得發抖。
4. 老太太坐著，四周圍繞著小孩子。
5. 那隻狗被埋在墓穴裡，躺在那裡。

寫作參考

1. Those girls came to the city singing all the way.
2. We sat on the top of mountain chatting and singing.
3. Many people stood in front of the church shivering with cold.
4. The old woman sat (being) surrounded by children.
5. The dog lies (being) buried in the grave.

▶ 知覺動詞（see、hear、feel...），若加了受詞之後，其後可搭配以下用法：

1. 原形動詞。
2. 依情況使用現在分詞或過去分詞。

Example：

我看到他與狼共舞。

I saw him dance with wolves.

Example：

我走進來時，看到他正與狼共舞。

I saw him dancing with the wolves when I walked in.

Example：

我今早看到他被狼群咬。

I saw him bitten by the wolves this morning.

Example：

我看到鳥群們在樹上跳躍。

I see the birds hopping on the tree.

Example：

我看到Tom正被老師處罰。

I saw Tom (being) punished by the teacher.

Example：

我在那時聽到她呼叫求救。

I heard her crying for help at that time.

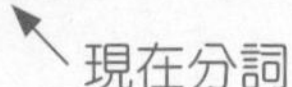

Example：

我覺得我身體內有東西在移動。

I felt something moving in my body.

現在分詞

Example：

我覺得我自己正被舉起來。

I felt myself (being) lifted.

being可省略

★ 特別說明

find之用法與以上感官動詞相似，可用現在分詞或過去分詞做受詞補語，但不可用原型動詞。

Example：

我發現他正與狼共舞。

I found him dancing with those wolves.

Example：

我發現這工作在匆忙中完成。

I found the work (being) done in a rush.

Example：

我發現這城市充滿了難民。

I found the city crowded with refugees.

▶ have、make可為使役動詞，換言之，用原型動詞做受詞補語。

Example:

我要求 Tom 洗那些衣服。

I made Tom wash the clothes.

▶ have、make亦可用過去分詞做受詞補語，以表被動。

Example:

我把我所有衣服都洗了。

I had all my clothes washed.

Example:

我對西班牙語一無所悉。

I can't make myself understood in Spanish.

▶ 在【make＋受詞＋過去分詞／原形動詞】的句型中，make之後的受詞，一般而言，為【人】而非【物】。但have及get則無此限制。

Example:

我將會要求他做那件事。

I'll make him do it.

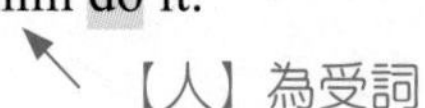

Example：

我將會洗那輛車。

I will make the car washed. (×)

正解：

I will have the car washed. (○)

I will get the car washed. (○)

寫作練習

1. 我們看到花園裡有許多玫瑰花盛開著。
2. 颱風來之前務必請人把窗戶修理一下。
3. 我不喜歡見到動物被殘酷地對待。
4. 他們正在尋找一隻小鳥，今晨他們曾聽到它正在樹上唱歌。

寫作參考

1. We saw many roses blooming in the garden.
   ↖ = In full bloom
2. Be sure to have the windows fixed before the typhoon comes.
3. I don't like to see animals cruelly treated.
4. They are looking for a little bird which they heard singing on the tree this morning.

寫作練習

1. 樓上有個人想要拜訪你。
2. John很怕他爸爸，他似乎對他很兇。
3. 信不信由你，他是個什麼都不怕的人。
4. 我已經不是你第一次認識的那個人。
5. 儀容邋遢的演講人幾乎得不到聽眾的注意。
6. 這是一則短篇小說，我很喜歡它那簡單的筆調。
7. 那個欠你一百元的人在哪裡。
8. 在四十位同學中，它是唯一考及格的人。
9. 他對人很友善，這一點大家都知道。
10. 這是我所能找到最令人信服的證據，以證明我的觀點。

寫作參考

1. There is a man upstairs who wants to call on you.
2. John is afraid of his father very much, who seems to be hard on him.
3. Believe it or not, he is the one who doesn't fear anything.
4. I am not the man that I was when you first knew me.
5. A speaker whose posture is sloppy can hardly get the attention of his listeners.
   = A speaker with sloppy posture can scarcely get the attention of his listeners.
6. This is a short story whose easy style I love very much.
7. Where is the man that owes you one hundred dollars?
8. Of the forty students, he is the only one that passed the exam.
9. He is friendly to others, which everyone knows.
   = As everyone knows that he is friendly with others.
10. This is the most convincing evidence that I can find to prove my point of view.

Chapter 2 動名詞（Gerund）

▶ **含動名詞之構句，如果表示不得不、不禁、忍不住，則有以下三種句型：**

1. Type A → 主詞 + cannot but + 原型動詞

Example：

每當我聽到這故事，我就忍不住哭泣。

Whenever I hear the story, I cannot but cry.

↖ 原型動詞

2. Type B → 主詞 + cannot help + 動名詞

Example：

每當我聽到這故事，我就忍不住哭泣。

Whenever I hear the story, I cannot help crying.

↖ 動名詞

3. Type C → 主詞 + cannot help but + 原型動詞

Example：

每當我聽到這故事，我就忍不住哭泣。

Whenever I hear the story, I cannot help but cry.

↖ 原型動詞

▶ **比較：在以下情況，若help當幫助之意，則以【人】做受詞，此時第二個動詞前要加to。**

Example：

由於Tom極為忙碌，我不得不幫他做那件事。

As Tom is extremely busy, I cannot help him to do the work.

↖第二個動詞前要加to

▲ cannot help it意指對某事無可奈何，在此it為代名詞，用以代替前面整個句子。因此，在此it不可為this或that取代。

• 易犯錯誤範例

Example：

他常說別人的不是，他知道這那是錯的，但他就是忍不住。

He often speaks ill of others; he knows this is wrong, but he just can't help that. (×)

He often speaks ill of others; he knows this is wrong, but he just can't help this. (×)

正解：

He often speaks ill of others; he knows this is wrong, but he just can't help it. (○)

▲ "there is no + Ving" 做……是不可能的

= It is impossible to + V

= No one can + V

Example:

未來十年會發生什麼事，是無法得知。

There is no telling what may happen in ten years.

= It is impossible to tell what may happen in ten years.

= No one can tell what may happen in ten years.

寫作練習

1. 當我獲知她的升遷時，我不得不佩服她。
2. 跟我老闆這種固執的人講理，簡直是不可能的事。
3. 實在無法形容這景色的美麗。
4. 得知她痛苦之遭遇時，我忍不住同情她。

寫作參考

1. Upon hearing her promotion, I cannot help but admire her.
2. It is impossible to reason with my boss, who is very stubborn.
3. No one can describe the beauty of this scenery.
4. Upon hearing her sufferings, I cannot help sympathizing her.

▶ Worth + 動名詞使用時機

worth爲介係詞，其後接名詞或動名詞爲受詞，形成介係詞片語，當形容詞用。

Example:

這件工作值得你的努力。

The work is worth your effort.

Example:

這本書值得你閱讀。

The book is worth reading.

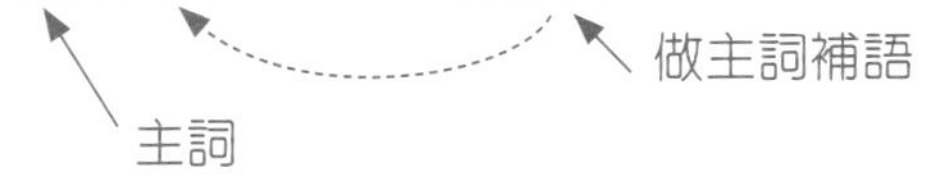

▶ 含有worth之句子，不可用虛主詞it，此為常見之錯誤用法。

• 易犯錯誤範例

Example:

這工作不值得做。

It is worth doing the work. (×)

正 解:

The work is worth doing. (○)

▶ 比較用法：worth / worthy

worthy為形容詞，使用時機為以下兩種情形：

1. be worthy of + 動名詞

2. be worthy to be + 過去分詞

Example：

這書值得閱讀。

The book is worthy of reading.

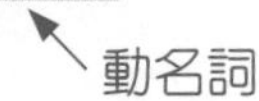

Example：

The book is worthy to be read.

▶ worthy和worth一樣，不可用虛主詞it做主詞。

• 易犯錯誤範例

Example：1

這書值得閱讀。

It is worthy of reading the book. (×)

正 解：

The book is worthy of reading. (○)

Example：2

這書值得閱讀。

It is worthy to read the book. (×)

正 解：

The book is worthy to be read. (○)

▶ 特別用法：worth while相當於形容詞，要用虛主詞it當主詞。

Example：

出國旅行是值得的。

It is worth while to take a trip abroad.

= It pays to take a trip abroad.

◎ 實用句型

no sooner... than 一……就……

= hardly... when

= scarcely... before

Example：

他一聽到這故事，就嚎啕大哭。

He had no sooner heard the story than he burst out crying.

= He had hardly heard the story when he burst out crying.

= He had scarcely heard the story before he burst out crying.

☆ **請注意**

(1)hardly和scarcely爲同義詞，故其後之連接詞when和before可互換。

(2)no sooner、hardly、scarcely均爲否定副詞，若其在句首出現，請切記要使用倒裝句寫法。

(3)倒裝句意味著動詞在前，動詞在後。

Example：

我一走，他就來。

I had no sooner left than he came.

或

No sooner had I left than he came.

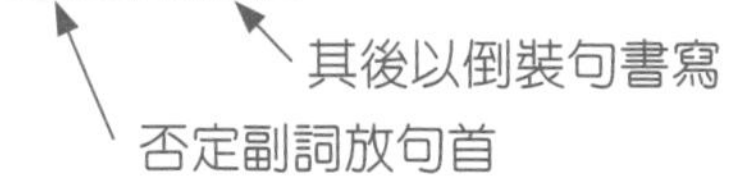

Example：

我一走，他就來。

I had hardly left when he came.

或

Hardly had I left when he came.

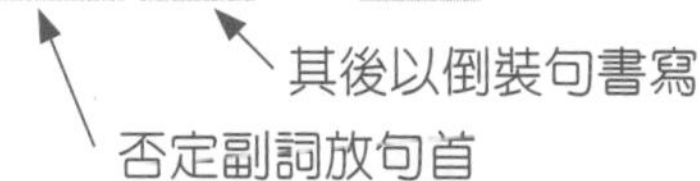

Example：

我一走，他就來。

I had scarcely left before he came.

或

Scarcely had I left before he came.

其後以倒裝句書寫

否定副詞放句首

▶ **表「一……就……」之句子，若副詞子句之主詞和主要子句之主詞相同時，可採下列句型：**

On + 動名詞 + that + 主詞 + 動詞

或

Upon + 動名詞，……主要子句

Example：

他一聽到那故事，就嚎啕大哭。

As soon as he heard the story, he burst out crying.

副詞子句　　主要子句

或

Upon hearing the story, he burst out crying.

動名詞

Example：

我一見到她，就哭了。

The moment I saw her, I cried.

或

Upon seeing her, I cried.

◎ 實用句型

It goes without saying that... 無庸置疑

= Needless to say

= It is needless to say

Example:

無庸置疑，文勝於武。

It goes without saying that the pen is mightier than the sword.

= Needless to say, the pen is mightier than the sword.

= It is needless to say that the pen is mightier than the sword.

◎ 實用句型

One who...　凡是……的人

Those who...　凡是…的人

Example:

凡是工作超時的人，在年終時都值得拿到紅利。

One who works over time is worth getting bonus at the end of year.

或

Those who work over time are worthy of getting bonus at the end of year.

寫作練習

1. 一知道秘書遲到，他的老闆就大發脾氣。
2. 不用說，健康勝於財富。
3. 我們的副總裁收到傳真時，連看都沒看就把它給扔了。
4. 值得一看的書，值得再看。

寫作參考

1. Upon hearing the late of the secretary, his boss lost his temper. (The boss had no sooner known that the secretary was late than he/she lost his/her temper)
2. Needless to say, the health is more important than the wealth. (It goes without saying that health is above wealth.)
3. Upon receiving the fax, the vice president threw it away without a glance.
4. A book which is worth reading is worth reading again. (A book which is worthy to be read is worthy reading again.)

◎ 實用句型

prevent...from + 動名詞　防止……／阻止……

Example:

疾病讓我無法上學。

Illness prevented me from going to school.

Example:

什麼原因讓他戒菸？

What stop him from smoking?

★ 延伸應用

prevent ban restrain prohibit stop keep hinder	+ 受詞 + from + 動名詞 = forbid + 受詞 + to

Example:

這大雨今早阻礙我們出門。

The heavy rain kept us from going out this morning.

Example:

這老師禁止他的學生玩電動遊戲。

The teacher bans his students from playing video games.

或

The teacher forbids his students to play video games.

☆ 請注意

forbid之受詞如果不是人，則直接用動名詞做受詞。

• 易犯錯誤範例

Example：

他們禁止在此處吸菸。

They forbid to smoke here. (×)

正 解：

They forbid smoking here. (○)

◎ 實用句型

encourage + 人 + to 鼓勵某人

discourage + 人 + from + 動名詞 打擊某人

Example：

他父親鼓勵他嘗試溜冰。

His father encourages him to try ice skating.

Example：

此失敗讓他不願再嘗試。

The failure discourage him from trying again.

◎ 實用句型

persuade + 人 + to 說服

dissuade + 人 + from + Ving 勸阻

Example:

這位母親說服她的小孩停止哭泣。

The mother persuaded her child to stop crying.。

Example:

他勸我不要抽菸。

He dissuaded me from smoking.

寫作練習

1. 年紀大，使他找不到新工作。
2. 無人能阻止Tom創立一個新公司。
3. 那父親禁止他女兒與Tom訂婚。
4. 任何困難都不能阻撓他嘗試。
5. 那件厚外套保護他，使他不被凍死。

寫作參考

1. Old age prevents him from getting a new job.
2. Nobody could keep Tom from setting up a new company.
3. The father forbids his daughter to be engaged to Tom.

4. No difficulty could discourage him from trying.
5. The thick coat protected him from being frozen to death.

◎ 實用句型

動名詞做受詞之使用時機：

That所引導之名詞子句，通常不能直接做介係詞之受詞，如果要做介係詞受詞，必須要變成動名詞片語。

◆ 步驟

A. 刪連接詞that

B. 子句中主詞變成所有格

C. 其後之動詞改成動名詞

• 易犯錯誤範例

Example：1

他工作如此認眞，我很好奇。

I am curious about that he works so hard. (×)

That所引導之名詞子句（that前不可有介係詞）

正 解：

I am curious about his working so hard. (○)

主詞變成所有格　　動詞改成動名詞

或

I am curious about the fact that he works so hard. (○)

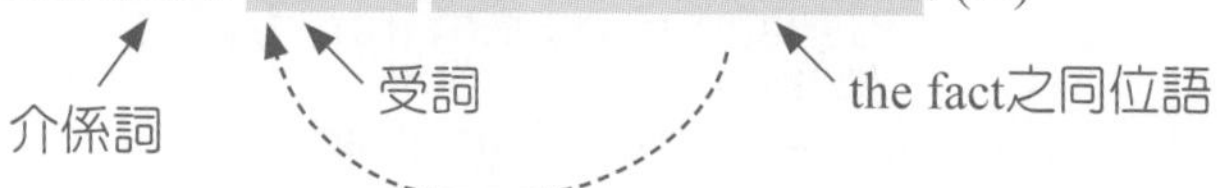

Example: 2

我很在乎他終日無所事事。

I am concerned about that he plays around all day. (×)

正 解:

I am concerned about his playing around all day. (○)

或

I am concerned about the fact that he plays around all day. (○)

▶ 認知動詞（例如：believe、think、find、know、feel、remember），其後常使用that接子句。

Example:

我相信他可勝任。

I believe that he can do it.

Example:

我認為明天就會沒事。

I think that it will be fine tomorrow.

▶ **意志動詞（例如：suggest、ask、order）表示〔建議〕、〔要求〕、〔命令〕、〔規定〕時，在其後，原則上使用助動詞should。而should往往加以省略，而直接使用原型動詞。**

Example：

這老板建議他應該努力工作。

The boss suggests that he (should) work hard.

Example：

他們要求他在5點前完成。

They asked that he (should) finish it by five.

Example：

他們命令他放下槍。

They ordered that he (should) drop the gun.

Example：

他們建議他安靜。

They suggest that he (should) be quiet.

寫作練習

1. 你記不記得五年前他搶了你的錢。
2. 他以他的兒子能當個好老師為榮。
3. Tom毫無希望能通過考試。
4. 就Peter說謊一事，我感到很訝異。
5. 這次失敗導致他離家出走。

寫作參考

1. Do you remember that he robbed you of your money five years ago?

   或

   Do you remember his robbing you of your money five years ago?
2. He is proud of his son's being a good teacher.

   或

   He is proud of the fact that his son is a good teacher.
3. There is no hope of Tom's passing the exam.

   或

   There is no hope of the fact that Tom will pass the examination.
4. I am surprised at Peter's telling lies.
5. The failure resulted in his running away from home.

▶ 但如遇到【喜歡】、【憤怒】等動詞，就不能直接用that子句，做為前面所提動詞之受詞。

以下爲將that子句變成動名詞。

• 易犯錯誤範例

Example：1

他的雙親不喜歡他與那女孩有牽連。

His parents don’t like that he associates with the girl. (×)

正 解：

His parents don’t like his associating with the girl. (○)

變成動名詞

或

His parents don’t like the fact that he associates with the girl. (○)

Example：2

我很享受她替我工作。

I enjoy that she does work for me. (×)

正 解：

I enjoy her doing work for me. (○)

或

I enjoy the fact that she does work for me. (○)

▶ 在寫作實務上，remember該動詞之後，可接that子句或動名詞片語，做受詞。

Example：

我記得我們10年前見過。

I remember that we met ten years ago.

或

I remember our meeting ten years ago.

Chapter 3 不定詞（Infinitive）

▶ 不定詞（Infinitive）用以表目的，結果之用法：

1. 表目的：

Example：

他大老遠來此看Mary。

He came all the way here to see Mary.

= in order to

Example：

他認真讀書以通過考試。

He studied hard to pass the examination.

= in order to

2. 表結果：

only to + 動詞　竟然（與預料相反之結果）

Example：

他認真讀書，考試竟然失敗。

He studied hard only to fail in the examination.

Example：

他沿路跑到車站，竟然沒趕上火車。

He rushed all the way to the station only to miss the train.

▲ 用以表【目的】之不定詞片語，可置於句首，但其後必須要有逗點。

Example：

他大老遠來此看Mary。

He came all the way here to see Mary.

或

To see Mary, he came all the way here.

Example：

他認眞讀書爲的是通過考試。

He studied hard to pass the examination.

To pass the examination, he studies hard.

▲ 活用寫作

Example：

他認眞學習，爲的是要通過考試。

He studied hard to pass the examination.

或

He studied hard in order to pass the examination.

或

He studied hard so as to pass the examination.

或

He studied hard with a view to passing the examination.

↖ 使用動名詞

或

He studied hard with an eye to passing the examination.

↖ 使用動名詞

▲ with a view to / with an eye to 為了

此處之to為介係詞（Prepositions），其後應使用動名詞。

Example:

為了改進他的英文，他訂閱了英文報紙。

With an eye to improving his English, he subscribes to an English newspaper.

寫作練習

1. Tom嘗試藉增加工作時數來提高他的工作績效，竟然令他的同事對他嫉妒。
2. 我們的秘書會說八種語言，一定是天才。
3. Mary必須在六點前起床，以免趕不上飛機。
4. Tom兼差當家教養家。
5. 聽到令堂臥病在床，我很難過。

寫作參考

1. Tom tries to increase his working hours to enhance his working performance only to make his colleague jealous of him.
2. Our secretary must be a genius to speak eight different languages.

3. Mary must get up by six o'clock so as not to miss the plane.

或

Mary must get up by six o'clock lest she should miss the plane.

4. Tom moonlights as a tutor to support his family.
5. I am sorry to hear that your mother is sick in bed.

▶ too...to... 太……因而不能……

Example:

他太年輕，因而不能做那個工作。

He is too young to do the work.

Example:

他走得太慢，因而不能趕上我。

He walked too slowly to catch up with me.

▶ 以上句型，常以下列句型取代：

so...that...cannot 如此……以致於不能……

Example:

他如此肥胖，以致於不能再行走。

He is too fat to walk any more.

= He is so fat that he cannot walk any more.

Example:

他手邊有太多工作，以致於不能和我們去野餐。

He has too much work on hand to go picnicking with us.

Example:

他有太多事要處理，以致於不能早一點回家。

He has too many things at his disposal to go home early.

特別用法1：

too...not to...　太……而令……

Example:

我非常高興而令我（願意）幫助那些男孩們。

I am too glad not to help the boys.

= I am very glad to help the boys.

= I am only too glad to help the boys.

特別用法2：

在too…to之句中，在too後方可接形容詞＋單數名詞。

- **易犯錯誤範例**

Example: 1

就開車而言，他還是個年輕之男孩。

He is a too young boy to drive a car. (×)

正解：

He is too young a boy to drive a car. (○)

形容詞 ↗ ↖ 單數名詞

Example：2

要駕駛那小船，這些男孩們還太年輕。

They are too young boys to steer the boat. (×)

↖ 應使用單數名詞

正解：

The boys are too young to steer the boat. (○)

▶ so...that...（如此……以致於）句中，亦可接含有形容詞之單數名詞，與too...to...句型相同。

Example：

他是如此年老的長者，以致於無法跳躍障礙。

He is so old a senior that he can't jump over the barriers.

↖ 含有形容詞之單數名詞

Example：

他們是如此年老的長者，以致於無法跳躍障礙。

The seniors are so old that they can't jump over the barriers.

▲ 在such...that...（如此……以致於）句中，such可視爲形容詞，其後可用不可數名詞、單數可數名詞、複數名詞。

A. 不可數名詞

Example：

這是如此好的音樂，以致於我很沉醉。

It is such good music that I enjoy it very much.

↖ 不可數名詞

B. 單數可數名詞

Example：

他是如此好的男孩，以致於我喜歡他。

He is such a nice boy that I like him.

↖ 單數可數名詞

C. 複數名詞

Example：

他們是如此好的男孩們，以致於我喜歡他們。

They are such nice boys that I like them.

↖ 複數名詞

寫作練習

1. 那部電影太無聊了，讓我看不下去。
2. 關於中了獎券一事，使得Tom笑得合不攏嘴。
3. 他似乎累得不能再工作下去。
4. 在Mary出國前，Tom沒有足夠的勇氣向他表達愛意，是一件很大的遺憾。
5. 她美得令我瘋狂。

寫作參考

1. The movie was too boring for me to see.

   或

   The movie was so boring that I can't see it.
2. Tom is too glad to stop laughing regarding winning the lottery.
3. It seems that he is too tired to work anymore.

   或

   It seems that he is so tired that he cannot work any longer.
4. It is a great pity that Tom didn't have enough courage to express his love toward Mary before she went abroad.
5. She is so pretty that I am crazy about her.

▶ 在疑問詞之後，加不定詞片語，可形成名詞片語。

Example:

如何開始比何處停止還困難。

How to begin is more difficult than where to stop.

Example:

請告訴我下一步如何做。

Tell me what to do next.

↖ 名詞片語

▲ how、where、when、why均爲疑問副詞，不須做其後不定詞片語中，動詞之受詞。

因此：

A. 若該動詞爲【及物動詞】，其後有受詞。

Example：

How to do it.

B. 若該動詞爲【不及物動詞】，其後無受詞。

Example：

Where to leave.

When to come.

Why to leave.

▲ what、whom、which爲疑問名詞，做其後不定詞片語中，動詞之受詞。否則，就做介係詞之受詞。

Example：

What to see.

Which to buy.

Whom to work with.

▶ **以下之名詞片語應視為名詞。因此，可做主詞、受詞、或be動詞之後的補語。**

1. 做主詞

Example：

如何做比何時做來的重要。

How to do it is more important than when to do it.

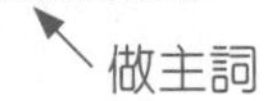

2. 做受詞

A. 做及物動詞之受詞

Example：

我不知何時寫那封信。

I don't know when to write the letter.

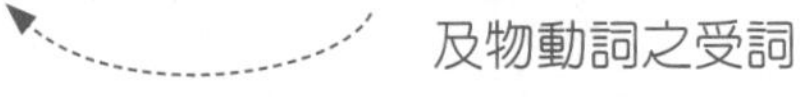

B. 做介係詞之受詞

Example：

有關要做什麼，我很茫然。

I am at loss about what to do.

3. 做be動詞之後的補語

Example：

問題是何時拿到我們需要的那筆錢。

The problem is when to get the money (which) we need.

be動詞後的補語

寫作練習

1. 我的老闆是這麼地固執，以致於我不敢告訴他如何以他個人密碼傳送那份傳真，尤其當他心情不好的時候。
2. 在展示間內有這麼多豪華進口車，以致於我完全沒有概念要買哪一輛。
3. 如何去瞭解我們的顧客需要什麼，比什麼時候和他們簽約來的重要。
4. 如何過活，對每一個人而言，是個重要的問題。

寫作參考

1. My boss is so stubborn that I dare not advise him on how to send the fax with his personal code especially when he is in bad mood.
2. There are so many luxurious imported cars on exhibition/display in the show room that I totally have no idea regarding which to buy.

   或

   There are too many luxurious imported cars on exhibition/display in the show room for me to make a decision due to the fact that I totally have no idea regarding which to buy.)
3. How to know what our customers want is more important than when to sign the contract.

4. How to live is an important question to everyone.

或

As far as everyone is concerned, how to live is an important question.

▶ **使役動詞（例如：make、have、bid，三個動詞之後，若加了受詞，以原形動詞做受詞補語）。**

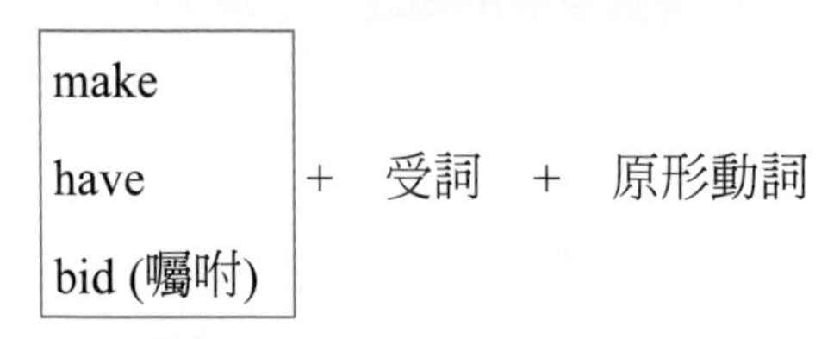

Example：

我要求他做這件事。

I made him do it.

原形動詞，當受詞補語

Example：

他們要求那男孩清理地板。

They had the boy clean the floor.

Example：

那位媽媽叮嚀那小孩自己注意規矩。

The mother bad the child behave himself.

▲ 一般而言，以下動詞之後，常用不定詞。

want、get、force、compel、expect、ask、would like、advise、urge、wish、tell

Example:

我告訴Mary別吃太多。

I told Mary not to eat too much.

Example:

我希望你保持緘默。

I want you to be quiet.

Example:

他們強迫我做那些違反我意願之事。

They forced me to do things against my will.

Example:

那老師力勸學生們認真學習。

The teacher urged students to work hard.

▲ hope該動詞後，只能接that子句做受詞。

• 易犯錯誤範例

Example:

我希望他能幫我做那件事。

I hope him to do it for me. (×)

正 解:

I hope that he could do it for me. (○)

▲want、expect、wish等動詞後，絕不可以用that子句做受詞。

A. want

• **易犯錯誤範例**

Example：1

我希望他工作認眞。

I want that he works hard. (×)

正 解：

I want him to work hard. (〇)

Example：2

我希望在那桌上有本書。

I want that there is a book on the desk. (×)

正 解：

I want there to be a book on the desk. (〇)

B. expect

Example：

我期待在那桌上有本書。

I expect there to be a book on the desk. (〇)

C. wish

Example：

她希望在她的領域中，是位專業經理人。

She wishes to be a professional manager in her fields. (〇)

▲ 有關enable動詞之用法：

enable + 受詞 + to

Example：

高科技讓我們享受舒適之生活。

High technology has enabled us to enjoy a comfortable life.

受詞

寫作練習

1. 幽默感使辦公室每個人能與他人相處愉快。
2. 我希望在我書桌旁能有個魚缸。
3. 繁重的工作迫使我取消去露營。
4. 他們幫助我，使我能夠順利完成工作。

寫作參考

1. A sense of humor enables everybody in the office to get along well with others.
2. I want there to be a fish tank by my desk.

   或

   I hope that there is a fish tank by my desk.
3. The heavy workload forced me to cancel the camping trip.

   或

The heavy-loaded work enabled me to abort the camping trip.

4. Their help enabled me to finish the work with ease.

▶ 獨立副詞片語之不定詞片語，用以修飾主要子句。通常放於句首，之後加逗點。

Example：

老實說，我不同意你。

To tell the truth, I don't agree with you.

↖獨立副詞片語　↖修飾主要子句

Example：

坦白說，我不喜歡你。

To be frank with you, I do not like you.

類似片語與用法

- To do (someone) justice 替（某人）說句公道話
- To be brief 簡而言之
- To begin with 開始
- To sum up 總結
- To make matter worse 讓事情更糟的是
- So to speak 換句話說
- To say nothing of 更不用說

Example:

替他說句公道話，他沒像你描述的那樣壞。

To do him justice, he is not as bad as you described.

Example:

簡而言之，我買不起它。

To be brief, I can’t afford to buy it.

Example:

學英文不難，開始時，你需要一本好字典。

It is not hard to learn English. To begin with, you need a good dictionary.

Example:

換句話說，Tom是本活字典。

Tom is, so to speak, a walking dictionary.

類似片語與用法

- have nothing to do with 與……毫無關係
- have something to do with 與……有點關係
- have much to do with 與……頗有關係
- have little to do with 與……幾乎無關

Example:

一個人的未來，與他/她的努力頗有關係。

One’s future has much to do with one’s effort.

Example:

這天氣和你的預言，毫無關係。

The weather has nothing to do with your prediction.

寫作練習

1. 說句公道話，她是個心腸很好的女孩子。
2. 他可說是個多才多藝的人。
3. 我勸你不要跟那個頑童有任何瓜葛。
4. 不用說，學而不思則罔。

寫作參考

1. To do her justice, she is a kind- hearted girl.
2. He is, so to speak, a versatile person.
3. I advise you not to have anything do with that naughty boy.
4. Needless to say, learning without thinking is useless.

   或

   Needless to say, it is no use learning without thinking.

Chapter 4 時態（Tense）

▶ 簡單現在式

1. 表不變眞理

Example：

夏季在春季之後。

Summer follows spring.

2. 習慣性之動作

Example：

我每天早上7點起床。

I get up at 7.

▶ 現在進行式

Example：

有人正在敲門，你可以應門嗎？

Someone is knocking the door. Can you answer it?

▶ 簡單過去式

Example：

Sam一個月前打過電話給我過。

Sam phoned me a moment ago.

▶ 過去進行式

1. 過去進行式：表示過去某時刻正在進行的動作或發生的情況。

Example：

我1988年時住在國外，所以我還蠻懷念臺灣。

I was living abroad in 1988, so I miss Taiwan.

Example：

正當我要離開屋子時，電話響了。

Just as I was leaving the house, the phone rang.

2. 並行的動作：可用while等強調同時進行的兩種或幾種動作。

Example：

當我正在花園工作時，Helen正在準備晚餐。

While I was working in the garden, Helen was cooking dinner.

▶ 簡單現在完成式

用以表示開始於過去，並持續到現在（或許會持續下去）的動作。

▲ since和for運用於現在完成式之情況：

→ since當做連接詞用時，其後可接簡單過去式或現在完成式。

→ 常有時間副詞（just、recently、already...）之搭配。

A. 當since爲連接詞時

Example：

自從還是個小男孩，就沒有回家過。

Tom has not been home since he was a boy.

連接詞

Example：

我在1980年退休後來到此處，就一直住在這裡。

I retired in 1980 and came to live here. I have lived here since I retired.

連接詞

Example：

幾年前，我已經住在這裡，而且自從我住在這裡，我認識很多朋友。

I have lived here several years and I have made many new friends since I lived here.

連接詞

B. 當since爲副詞時

Example：

我在5月見過Mary，自那時開始，就沒有再見過她。

I saw Mary in May and I have not seen her since.

副詞

C. 當since爲介詞時

Example：

我1980年起就住在這裡。

I have lived here since 1980.

介詞

☆ 請注意

for雖常與現在完成式連用，但也可以和其他時態連用。

→ 須注意語意

Example：

我住在那裡已有5年。

I have lived there for five years. 意謂著 → 現在仍住在那裡。

Example：

我曾住在那裡5年

I lived there for five years. 意謂著 → 現在不住在那裡。

▶ 簡單過去完成式

had + 過去分詞

主要用法表示：兩個事件何者發生在前

Example：

當醫師抵達時，那病人剛過世。

The patient died when the doctor arrived

Example：

當醫師抵達時，那病人已過世。

The patient had died when doctor arrived.

▶ 現在完成進行式／過去完成進行式

1. 現在完成進行式: **have been + Ving**

   強調動作在某一段時間內，一直在進行，而且這動作往往對現在產生結果。

Example:

她非常疲倦，她今天到現在仍在打書信。

She is very tired. She has been typing letters all day.

2. 過去完成進行式：had been + Ving

A. 表示比過去更早一段時間內進行的動作，而且往往對過去某一時刻產生結果。

Example:

她過去非常疲倦，她過去一直持續在打書信。

She was very tired. She had been typing letters all day.

B. 表示重複動作的現在／過去完成進行式

Example:

Jim從過去幾個星期到現在，每晚打電話給Jenny。

Jim has been phoning Jenny every night for the past weeks.

Example:

Jenny被打擾了，（因為）Jim在過去一星期每晚打電話給她。

Jenny was annoyed. Jim had been phoning her every night for a whole week.

比較：

Example:

當我到家時，我發現Jill還一直在粉刷她的房間。

When I got home, I found that Jill had been painting her room.

Example:

當我到家時，我發現Jill已粉刷完她的房間。

When I got home, I found that Jill had painted her room.

▶ 簡單將來式

will / shall + 原型V

表示將來的其他方法：

1. be going to

Example:

我將打電話給我老闆。

I am going to call my boss.

2. be to

Example:

我將見到Tom。

I am to see Tom.

▶ 將來進行式

will / shall + be + Ving（現在分詞）

表示將來進行著的動作。

Example:

快點，客人們隨時都會到來。

Hurry up, the guests will be arriving at any minutes.

▶ 簡單將來完成式

will have + p.p.（過去分詞）

用於表示到將來某一時刻，已經完成的動作。

Example:

公元2020年以前我會完成退休。

I will have retired by the year 2020.

Example:

我期待明天之前，你會改變你的看法。

I expect you will have changed your mind by tomorrow.

▶ 將來完成進行式

will have been + Ving

Example:

下星期的此時，我將爲這公司工作滿24年。

By this time next week, I will have been working for this company for 24 years.

寫作練習

1. 他現在正在工作，所以不能接LINE。
2. 我進來時，Jim在和他女朋友通電話，一小時之後我出去時，他還在跟她通電話。
3. 男孩們喜歡動物園，以前他們從來沒有見過野生動物。
4. 她的眼睛紅了，顯然她哭過了。
5. 五年後，將有一個太空飛行器繞木星飛行。
6. 到12月25日，我們倆結婚就滿一年了。
7. Tom一開始講話，就被人打斷。

寫作參考

1. He is working at the moment, so he can't answer the LINE.
2. Jim was talking to his girlfriend on the phone when I came in and was still talking to her when I went out an hour later.
3. The boys loved the zoo. They had never seen wild animals before.
4. Her eyes are (were) red, it is obvious that she has cried.
5. A space vehicle will be circling Jupiter after five years.
6. We will have been married a year on December 25th.
7. Tom had hardly begun his speech when he was interrupted.

   或

   Tom had no sooner begun his speech than he was interrupted.

   或

   Tom had hardly begun his speech when he was interrupted.

   或

   Tom had scarcely begun his speech before he was interrupted.

   或

   As soon as he began his speech, he was interrupted.

   或

   Upon beginning his speech, he was interrupted.

或

On beginning his speech, he was interrupted.

或

The moment he began his speech, he was interrupted.

▶ 假設語法（Subjunctive）

共有四種類型：

1. 純條件 (A)
2. 與現在事實相反 (B)
3. 與過去事實相反 (C)
4. 與未來狀況相反 (D)

A. 純條件

(1)使用時機為不知實際狀況為何，但卻只要某條件成立，則必然有預期之結果。

(2)假設子句用現在式，主要子句用現在式助動詞，加原形動詞。

格式如下：

If...現在式動詞……，主詞 + will / shall / may / can / must / ought to / should + 原形動詞

☆ 請注意

should = ought to表【應爲】之意，是唯一可在純條件假設中使用之過去式助動詞，would、might、could就不可使用。

Example：

如果我富有，我會買輛車。

If I am rich, I will buy a car.

Example：

如果我有錢，我也許會多捐點。

If I have money, I may donate more.

Example：

如果Tom想要通過考試，他必須認真工作。

If Tom desires to pass the examination, he must work hard.

Example：

如果你見到她，你應該告訴她事實。

If you see her, you should tell her the truth.

• 易犯錯誤範例

Example：1

如果我有錢，我會買房。

If I am rich, I would buy a house. (×)

↖應使用現在式助動詞

正 解：

If I am rich, I will buy a house. (○)

Example: 2

如果我有錢，我或許會給他一些。

If I will have money, I may give him some. (×)

↖ 假設子句應使用現在式

正 解:

If I have money, I may give him some. (○)

▲ unless亦可當作純條件假設語氣使用（unless = if）

Example:

如果我有錢，我會買房。

If I am rich, I will buy a house.

= Unless I am rich, I will not buy a house.

▲ If 可被provided (that)取代

Example:

如果我有時間，我會拜訪他。

If I have time, I will call on him.

或

Provided (that) I have time, I will call on him.

或

In case I have time, I will call on him.

或

Suppose I have time, I will call on him.

或

Supposing I have time, I will call on him.

或

I will call on him on condition that I have time.

寫作練習

1. 除非你記下來，要不然當你忘了重要的會議記錄時就太晚了。
2. 如果你的頭期款是現金，我會和你簽汽車保險約。
3. 如果你愛那年輕人並要他保住性命的話，就要他放棄冒險。
4. 今天下午我有空的話，就跟你們踏青去。
5. 如果畫完成了，先讓我看一看。
6. 如果你肯幫我們搬這箱子，我們將感激不盡。

寫作參考

1. Unless you take a note, it will be too late that you will forget important minutes.
2. If your down payment is cash, I will sign the contract with you regarding your car insurance.
3. If you love that young man and want to keep him alive, you should make him give up the adventure.
4. If I am free this afternoon, I will go hiking with you.

5. If the picture is completed, let me take a look first.
6. Provided (that) you help us move the box, we will be very appreciated.

B. 與現在事實相反

格式如下：

If...過去式動詞……，主詞 + could / might / should / would / ought to + 原形動詞

Example：

如果我知道他的住址，我就能寫信給他。

If I knew his address, I could write to him.

→意味著與現在事實相反，換言之，我現在並不知道他的住址。

Example：

如果我有望遠鏡，我就能很清楚看見那慧星。

If I had binoculars, I could see that comet clearly.

→意味著與現在事實相反，換言之，我現在並沒有望遠鏡。

Example：

如果我是一隻鳥，我就能飛向你。

If I were a bird, I would fly to you.

↖過去式動詞

→意味著與現在事實相反，換言之，我現在並不是一隻鳥。

▲ 表示與現在事實相反之be動詞，不論主詞爲第幾人稱，都使用were。

Example：

如果他身體狀況好，他會出國。

If he were in good health, he would go abroad.

↖ 使用were

→意味著與現在事實相反，換言之，他現在身體並不好。

▲ were可置於主詞前，if可替換爲Provided (that)。

Example：

如果你在我的處境，你會怎麼做？

If you were in my shoes, what would you do?

= Provided (that) you were in my shoes, what would you do?

= Were you in my shoes, what would you do?

↖ were置於主詞前

▲ It is time + 過去式that子句

that子句用過去式之理由：爲了表示與現在事實相反。

Example：

是你該前往執勤的時候了。

It is time that you left for the duty.

↖ 使用過去式

比較：

Example：

我正要離開台北。

I am leaving Taipei.

Example：

我正要前往台北。

I am leaving for Taipei.

Example：

是他該出現的時候了。

It is time that he were here.

寫作練習

1. Mary已搬到美國去了，如果她現在在這兒，她會很高興跟我們在一起。
2. 我快五十了，如果我還年輕，我會多學一些外語。
3. 如果現在我老闆在此，他有義務幫我解決這問題。
4. 我要是（現在）有錢，我不會過那麼奢侈的生活。
5. 該是你反省思過的時候了。

寫作參考

1. Mary has moved to the United States. If she were here, she would be glad to be with us.
2. I shall be fifty. If I were young, I would learn a few more foreign languages.
3. If my boss were here, he would be obliged to help me solve the problem.

If my boss were here, he would have the obligation to help me solve the problem.

4. If I were rich, I will not live such a luxurious life.
5. It is time that you reflected upon your mistakes.

C. 與過去事實相反

格式如下：

(1)If子句用過去完成式，主要子句用過去式助動詞 + have + 過去分詞

(2)If... had + 過去分詞，主詞 + would / might / could / should / ought to + have + 過去分詞

Example：

如果我事先知道您飛機離境時間，我們就會在機場給您送行。

If we had known your departure, we would have seen you off at the airport.

had known ↖ 使用過去完成式

would have seen ↖ 使用過去式助動詞 + have + 過去分詞

Example：

如果他們有時間，他們會幫你。

If they had had time, they would have helped you.

had had ↖ 使用過去完成式

would have helped ↖ 使用過去式助動詞 + have + 過去分詞

▲ If句中had，可移到主詞前，而省略If。

Example：

如果他買了它，他會覺得後悔。

If he had bought it, he would have felt sorry.

= Had he bought it, he would have felt sorry.

移到主詞前

▲ 時態不一致時的假設語氣使用法：

If子句若與過去事實相反，用過去完成式表示。而主要子句若與現在事實相反，則用過去式助動詞表示。此類主要子句之句尾，一般而言，多有表現在的時間副詞。例如：now、today等。

Example：

如果他年輕時有努力工作，他現在應該很好。

If he had worked harder when he was young, he would be well of now.

Example：

如果10年前我很富有，我今天就會買那房子。

If I had been rich ten years ago, I could buy that house today.

寫作練習

1. 如果他當時說實話，現在就不會受處罰了。
2. 要是Tom當年向你求婚，你會答應嫁給Tom嗎？
3. 當時我要是知道他的困境，一定會幫助他的。
4. 當時他要是聽了醫生的建議，現在可能還活著。
5. 我當時要能預見這些困難，我就不會承辦這工作了。

寫作參考

1. If he had told the truth then, he would not be punished now.
2. If Tom had proposed to you, would you have married him?
3. If I had known his misery, I would certainly have helped him.
4. Had he followed the doctor's advice, he might be alive now.
5. If I had foreseen these difficulties, I wouldn't have undertaken the work.

D. 與未來狀況相反

(1) 與未來狀況相反之假設語氣，使用方式為：If子句要用助動詞should（萬一）。

(2) 主要子句之助動詞則視可能性做變化，若可能性低，則助動詞為過去式，若可能性高，則助動詞為現在式。

a. 可能性低

格式如下：

If...should...，主詞 + might / would / could / should / ought to + 原形動詞

Example：

萬一他來了，我會告訴他事實。

If he should come, I would tell him the truth.

使用過去式助動詞，表可能性低

b. 可能性高

格式如下：

If...should...，主詞 + shall / will / may / can / ought to / should + 原形動詞

Example：

If he should come, I will tell him the truth.

使用原形助動詞，表可能性高

▲ If子句中之should可置於主詞前，可將if省略。

Example：

如果那男孩再來，我會把他攆出去。

If the boy should come again, I would throw him out.

= Should the boy come again, I would throw him out.

☆ 請注意

if子句中，如有were、should，及過去完成式助動詞（had）出現時，均可將這些字，置於主詞前，將if省略。

(1) 有were（是）時

Example：

如果我是你，我不會做那件事。

If I were you, I wouldn't do it.

= Were I you, I wouldn't do it.

置於主詞前

(2) 有should（萬一）時

Example：

萬一下雨，我們的計畫就泡湯了。

If it should rain, our plan would be spoiled.

= Should it rain, our plan would be spoiled.

Example：

如果他有來，他就會見過 Mary.

If he had come, he would have seen Mary.

= Had he come, he would have seen Mary.

▲ If...were to與if...should用法接近，均表與未來狀況相反之假設語氣。但if...were to所表示之可能性更低，通常爲表示與眞理相反之假設語氣。

比較：

Example：

萬一他來了，我會告訴他。

If he should come, I would tell him. → 意謂在未來，他來之機率不高。

Example：

如果他眞來了，我會告訴他。

If he were to come, I would tell him. → 意謂在未來，他來之機率更低。

If the sun were to rise in the west, I would pass the examination.

▲ What if... should... 萬一……怎麼辦？

Example：

萬一他眞的回來，那怎麼辦？

What if he should come back?

= What would happen if he should come back?

= What would I do if he should come back?

寫作練習

1. 我要是再活一次，我想當一隻鳥。
2. 萬一太陽消失了，地球會變成什麼樣子？
3. 萬一另一次世界大戰爆發，人類將會發生什麼事？
4. 萬一發生地震，該怎麼辦？

寫作參考

1. If I were to be born again, I would like to be a bird.

   或

   Were I to be born again, I would like to be a bird.
2. If the sun were to disappear, what would the earth be like?
3. If another World War should break out, what would become of human beings?

   或

If another World War should break out, what would happen to human beings?

4. What could we do if an earthquake should take place?

Chapter 5 比較級與最高級（Comparative & Superlative）

▶ as...as... 和……一樣

1. 第一個as爲副詞，翻譯成【一樣地】。
2. 第二個as則爲副詞連接詞，翻譯成【和……】，引導副詞子句，修飾第一個as。

☆ 請注意

(1) 第一個as爲副詞，其後可爲形容詞或副詞。

(2) as...as只能與單數名詞並用，用法與so...that...一樣。

(3) 格式：as + 形容詞 + a/an + N + as

Example:

她像Mary一樣漂亮。

She is as beautiful as Mary (is).

Example:

他工作像 Peter 一樣認眞。

He works as hard as Peter (does).

↖ 副詞

• 易犯錯誤範例

Example：1

他像Peter一樣，是個好男孩。

He is as a nice boy as Peter is. (×)

正 解：

He is as nice a boy as Peter is. (○)

↖ 形容詞 + a + 名詞

Example：2

他們像Peter和David一樣，是好男孩。

They are as nice boys as Peter and David. (×)

↖ as...as 只能與單數名詞並用

正 解：

They are nice boys just like Peter and David. (○)

補充

- as many + 複數名詞 + as ... 像……一樣眾多……
- as few + 複數名詞 + as ... 像……一樣少……
- as much + 不可數名詞 + as ... 像……一樣眾多……
- as little + 不可數名詞 + as ... 像……一樣少……

Example：

她擁有的朋友像John一樣多。

She has as many friends as John (does).

Example：

她擁有的金錢像John一樣多。

She has as much money as John (does).

☆ 請注意

as...as可用於肯定或否定，而so...as僅可用於否定句中。

Example：

她不像Mary一樣美麗。

She isn't as beautiful as Mary (is).

或

She isn't so beautiful as Mary (is).

• 易犯錯誤範例

Example：

她像Mary一樣美麗。

She is so beautiful as Mary (is). (×)

正 解：

She is as beautiful as Mary (is). (○)

▶ as形容詞／副詞as one can = as形容詞／副詞as possible as one can 竭盡某人所能

Example：

我會竭盡所能小心開車。

I will drive a car as carefully as I can.

↖ 副詞

或

I will drive a car as possible as I can.

Example：

他竭盡所能早起，以趕上第一班車。

He got up as early as possible to catch the first train.

或

He got up as early as he could to catch the first train.

▶ as + 形容詞 + as形容詞can be 極……、不亞於任何人

Example：

他極爲英俊。

He is as handsome as (handsome) can be.

Example：

John工作極爲認眞。

John is as hardworking as (hardworking) can be.

Example：

Jane工作認眞不亞於班上任何人。

Jane is as hardworking as anyone in class.

▶ as + 形容詞／副詞 + as ever 與往常一樣

Example：

他似乎與往常一樣忙碌。

He seems to be as busy as ever.

Example：

他似乎與往常一樣認眞工作。

He works as hard as ever.

▶ as many用以代替前面提過之相同數字，以避免重複。

Example：

他在5個月閱讀5本書。

He read five books in five months.

→ He read five books in as many months.

Example：

他做10次犯10個錯誤。

He made ten mistakes in as many times.

寫作練習

1. 花旗銀行的出納正在下午三點半之前，忙著工作，就像蜜蜂一樣忙。
2. 當Mary看見她的老闆右手拿著CD向她衝過來時，她的臉色就跟紙一樣慘白。
3. 要想把英文講得很流利，你最好盡量多看英文文章。
4. 我沒料到他居然在三天內，犯了三次搶劫。
5. 他也許不如Mary聰明，但卻用功極了。
6. 每次我見到這女孩，我就會想起她的姐姐，她們長得一模一樣。

寫作參考

1. The cashiers of Citibank are as busy with their work as bees by 3:30pm.
2. When Mary saw her boss rushing toward her with a CD in his/her right hand, her face turned as pale as a sheet.
3. To speak English fluently, you had better read as many English articles as you can.
4. I didn't expect that he should have committed three robberies in as many days.
5. He may be not as clever as Mary, but he is as industrious as can be.
6. Whenever I see the girl, I am reminded of her sister. They are as like as two peas.

▶ ...times as...as...

1. 倍數詞 + as...as... 是……的幾倍

Example:

他學習認眞程度是我的3倍。

He studies three times as hard as I.

Example：

他的勤勞是我的3倍。

He is three times as diligent as I.

Example：

我的年紀是你的3倍。

I am three times as old as you.

2. more than + 倍數詞 + as...as... 是……的幾倍還不止

Example：

他學習認真程度是我的3倍以上。

He studies more than three times as hard as I.

Example：

我的年紀是你的3倍以上。

I am more than three times as old as you.

▲ 倍數也可以下列方式表示：

One – third → 1 / 3

Two – thirds → 2 / 3

Three – fourths → 3 / 4

Twice → 2

Three times → 3

Four times → 4

…

Example:

我的年紀是你一半。

I am half as old as you.

Example:

我學習認眞程度僅是Tom的三分之一。

I study only one-third as hard as Tom.

3. 倍數詞 + the（或所有格）+ 名詞　是……的幾倍

Example:

他擁有的錢是我的2倍。

He has twice my money.

Example:

這條河的長度是淡水河的10倍。

The river is ten times the length of Tamsui River.

或

The river is ten times as long as the Tamsui River.

4. more than + 倍數詞 + the（或所有格）+ 名詞　是……的幾倍還不止

Example:

他擁有的錢是我的2倍還不止。

He has more than twice my money

Example:

這條河的長度是淡水河的10倍還不止。

The river is more than ten times the length of the Tamsui River.

▶ as many as + 數字 + 複數N　多達……

Example:

多達100人在那場空難中喪生。

As many as 100 people were killed in the air crash.

Example:

多達5個蘋果在桌上。

There are as many as five apples on the table.

寫作練習

1. New Zealand之面積，大約是Taiwan的20倍。
2. 我每次聽Tom說英語時，真希望我能說得有他一半好。
3. Holland的居民是New York State的三分之二，而New York State 是Holland的四倍大。
4. 建築費用漲到高達新台幣兩千萬。
5. 他蒐集的郵票是我的三倍還不止。

寫作參考

1. New Zealand is about 20 times as large as Taiwan.

   或

   New Zealand is about 20 times the size of Taiwan.
2. Whenever I hear Tom speaks English, I wish I could speak

half as well as he.

3. Holland has two-thirds the residents of the state of New York, which is four times the size of Holland.

   或

   Holland has two-thirds as many inhabitants as the state of New York, which is four times as large as Holland.

4. Building cost ran up to as much as NT$20 million.

5. He has more than three times the stamps (which) I have collected.

   = He has collected more than three times as many stamps as I (have).

▶ The + 比較級……，the + 比較級……　越……越……

Example:

Tom 年紀越大，記憶力越差。

The older Tom grew, the weaker his memory becamc.

Example:

天氣越冷，我越覺得舒適。

The colder the weather (is), the more comfortable I feel.

Example:

你學習越認真，越會成爲更好的學生。

The harder you study, the better student you'll become.

Example:

你越小心，犯的錯越少。

The more careful you are, the fewer mistakes you'll make.

☆ 請注意

(1) 在此句型中，若句中無形容詞或副詞，則在the之後加more或 less。

Example:

這男孩越好，我就更喜歡他。

The better the boy (is), the more I like him.

Example:

你越好，我就更喜歡你。

The better you are, the more I like you.

(2) 若句中主詞爲一般名詞，而非代名詞（he、it、you、they、或專有名詞），若其後爲be動詞時，該be動詞可省略。

Example:

Tom越好，我就更喜歡他。

The better Tom is, the more I like him.

Example：

你越愛我，我對你越好。

The more you love me, the nicer I will be to you.

▶ 比較級 + and + 比較級…… 越來越……

Example：

這女孩變得越來越美麗。

The girl became more and more beautiful.

Example：

你應該越來越認眞學習。

You should study harder and harder.

Example：

在春季，天氣變得越來越溫暖。

It is getting warmer and warmer in spring.

▶ None the + 比較級 + because 即使……卻一點也不
None the + 比較級 + for + 所有格 即使……卻一點也不

Example：

即使他運動，他的健康卻一點也沒更好。

His health is none the better because he takes exercise.

His health is none the better for his taking exercise.

Example：

即使她穿時尚的洋裝，她卻一點也不美麗。

She is none the more beautiful because she wears a fancy dress.

She is none the more beautiful for her wearing a fancy dress.

寫作練習

1. 隨著工業迅速的發展，科學工業園區在國際舞台上的地位變得越來越重要。
2. 即使他有錢，他卻一點也不快樂。
3. 你越接近大自然，就越能了解它的美。
4. 你越努力，就越可能成功。

寫作參考

1. With the rapid industrial developments, the Science Based Industrial Park's position is getting more and more important on global stages.
2. He is none the happier for his great wealth.

   或

   He is none the happier because he has a lot of money.
3. The closer you stay to great nature, the more you will appreciate its' beauty.

或

The more you expose yourself to great nature, the more you will appreciate Its' beauty.

4. The harder you work, the more likely you are to achieve to success.

▶ No more...than　不是……正如……不是

Example:

鯨魚不是魚正如馬不是魚。

A whale is no more a fish than a horse is.

▶ No less...than　和……一樣

Example:

他和他哥哥一樣聰明。

He is no less clever than his elder brother.

相當於...as ... as...

He is as cleaver as his elder brother.

▶ not less ...than　至少與……一樣

Example:

她至少與Mary一樣美麗。

She is not less beautiful than Mary (is).

▶ ...no more than + 數字詞 + N = ... only + 數字詞 + N 僅僅

Example:

自台積電公司走到科學園區大門僅僅10分鐘。

It is no more than ten minutes walk from tsmc Inc. to the main gate of the Science-Based Industrial Park.

或

It is only ten minutes walk from tsmc Inc. to the main gate of Science-Based Industrial Park.

▶ ...no less than + 數字詞 + N = ... as many as + 數字詞 + N 剛好

Example:

自台積電公司走到科學園區大門剛好10分鐘。

It is no less than ten minutes walk from tsmc Inc. to the main gate of the Science-Based Industrial Park.

或

It is as many as ten minutes walk from tsmc Inc. to the main gate of the Science-Based Industrial Park.

▶ ...not more than + 數字詞 + N = ... at most + 數字詞 + N 最多不超過

Example:

自台積電公司走到科學園區大門最多不超過10分鐘。

It is not more than ten minutes walk from tsmc Inc. to the main gate of the Science-Based Industrial Park.

或

It is at most ten minutes walk from tsmc Inc. to the main gate of Science Based Industrial Park.

▶ ...not less than + 數字詞 + N = ...at least + 數字詞 + N 至少

Example:

自台積電公司走到科學園區大門至少10分鐘。

It is not less than ten minutes walk from tsmc Inc. to the main gate of the Science-Based Industrial Park.

或

It is at least ten minutes walk from tsmc Inc. to the main gate of the Science Based-Industrial Park.

寫作練習

1. 他不體諒別人，就像他弟弟一樣。
2. 剛剛好有十位同學在這次考試中被當了。
3. 由於老闆的惡意態度，至少有十名員工將會在拿年終獎金之前離職。
4. 他和瑪莉一樣擅長游泳。

寫作參考

1. He is no more considerate of others than his younger brother is.
2. No less than ten classmates failed the exam.
3. Not less than ten employees will quit their jobs before getting the bonus due to the boss's bad attitude.
4. He is as good at swimming as Mary.

   或

   He is no less good at swimming than Mary.

代名詞（Pronoun）

▶ one、we、you在某些時候使用時，泛指全體。

Example:

我們不該講別人之壞話。

One should not speak ill of others.

We should not speak ill of others.

You should not speak ill of others.

Example:

每個人都不該浪費自己的青春。

One should not waste one's youth.

We should not waste our youth.

You should not waste your youth.

▶ They say that...、people say that...、it is said that... 在某些時候使用時，泛指全體。

Example:

人們說他的父親非常富有。

They say that his father is very rich.

People say that his father is very rich.

It is said that his father is very rich.

▲ that of、those of 可做爲避免前、後重複之代名詞。

Example：

東京的人口比倫敦多。

The population of Tokyo is larger than that of London.

= the population

Example：

這些學生比我們學校的學生還認眞。

The students work harder than those of our school.

= the students

▲ 英文構句中，兩個名詞形成比較的情況時，爲避免重複，第二個名詞若爲單數，就改成代名詞that；若爲複數，則改成代名詞those。

Example：

在台灣的人民過著比柬埔寨人民更好的生活。

People in Taiwan lead a much richer life than people of the Cambodia.

→People in Taiwan lead a much richer life than that of the Cambodia.

that取代people

Example：

這裡的氣侯和臺中相仿。

The climate here is like the climate of Taichung.

→ The climate here is like that of Taichung.

寫作練習

1. 你不該違背良心做事。
2. 今秋的服裝款式，似乎與去年秋天大不同。
3. 雖然女性求職機會增加了，但男女的地位仍有很大的差距。
4. 這兩輛車子的顏色，我很難區分。
5. 她的孩子很有教養，但她姊姊的孩子卻調皮得很。
6. 古時候的人日子過得很苦，而現代人則因為機器的發明，生活過得舒適多了。

寫作參考

1. You should not do anything against conscience.
2. It seems that this autumn's fashions are quite different from those of last autumn.
3. Although the opportunities of employment of women are increasing, there is still a wide gap between the position of men and those of women.
4. It is very hard for me to distinguish the color of this car from that of that one.
5. Her children are well-educated, whereas her sister's kids are naughtier than those.

6. People in ancient time lived a hard life while those today enjoy a much more comfortable life because of the innovation of machines.

▲ the former... the latter... 前者……後者……

Example:

John和Peter是好朋友。前者是位老師，後者是位軍人。

John and Peter are good friends. The former is a teacher, the latter a soldier.

▲ the former... the latter... 句型敘述，可替換爲以下表示：

the former... the latter...

= that... this...

= the one... the other...

Example:

美德和罪惡是二件不同的事，前者帶來和平，後者導致悲慘。

Virtue and vice are two different things; the former leads to peace, the latter to misery.

(the former → 前者; the latter → 後者)

或

Virtue and vice are two different things; that leads to peace, this to misery.

(that → 前者; this → 後者)

或

Virtue and vice are two different things; the one leads to peace,

前者

the other to misery.

後者

☆ **請注意**

(1) the former、the latter可代表單、複數名詞；that、this和the one、the other只可代表單數名詞。

Example：

人類和機器人不同之處在於前者會笑，然而後者不會。

A man differs from robots in that the former is able to laugh, while

在此in that為副詞連接詞
= because the latter aren't.

(2) 若兩個名詞均為複數時，亦可用 those…these / the former…the latter 取代。

Example：

以某些觀點而言，狗比貓更忠實。貓被不同地方吸引著，狗則跟隨人類。

Dogs are more faithful animals than cats in some aspects; these attach

themselves to various places, and those to persons.

= Dogs

寫作練習

1. 健康勝於財富，換言之，後者無前者重要。
2. David和Mike是我的好友，前者喜歡唱歌，後者則迷戀攝影。
3. 愛情和麵包都很重要，前者充實我的精神生活，後者則充實我的物質生活。

寫作參考

1. Health is above wealth; that is to say, this is not so important as that.
2. David and Mike are both my good friends; the former enjoy singing, while the latter is obsessed with photography.
3. Love and bread are both important; the one enriches my spiritual life, while the other supplies my material life.

▲ One... the other...（僅適用於兩者）

Example:

我有兩位叔叔，一位住在東京市，另一位住在紐約市。

I have two uncles; one lives in Tokyo and the other lives in NYC.

▲ Some... others...（適用於多者）

Example：

有些人信上帝，有些人不信。

Some people believe in God and others don't.

▲ One... another... the other...（適用於三者）

Example：

他有三位兄弟，一位是老師，另一位是軍人，另一位是音樂家。

He has three brothers; one is a teacher, another (is) a solider, and the other is a musician.

▲ 比較用法：

Example：

我不喜歡這個，請給我看另一個。

I don't like this one; please show me another.

↖ 意味著展示品至少有3個

I don't like this one; please show me the other.

▲ 活用語法1：...is one thing, and... is another

Example：

知道是一回事，教是另一回事。

To know is one thing, and to teach is another.

▲ 活用語法2：One... the others

Example：

在我班上有40位學生，僅一位通過考試，其餘都被當掉。

I have 40 students in my class. Only one passed the exam, and the others all failed.

= (the rest)

寫作練習

1. 這兩個兄弟時常吵架，一個很固執，另一個則很自私。
2. 他們三人彼此相處愉快，一個已婚，一個人是光棍，而另一個則有了女友。
3. 賺錢是一回事，而花錢又是一回事。

寫作參考

1. The two brothers often quarrel with each other; one is stubborn and the other is selfish.
2. They three get along well with one another; one is married, another is still bachelor, and the other has a girlfriend.
3. To make money is one thing and to spend it is another.

連接詞（Conjunction）

▶ 在命令句中（也就是以原形動詞起首）的連接詞有兩種：

1. and表示一致的概念，中文詮釋爲【那麼】。

Example:

認眞點，那麼你就會通過考試。

Study hard, and you will pass the examination.

2. or表示相反的概念，中文詮釋爲【否則】。

Example:

認眞點，否則你無法通過考試。

Study hard, or you will failed (in) the exam.

▲ otherwise亦可取代or，但otherwise通常只做連接性副詞（有連接詞意味，但不見得有連接詞之功能），所以其前不用逗點，而用分號。

Example:

現在就戒菸，否則你的健康會被毀掉。

Stop smoking now; otherwise your health will be ruined.

寫作練習

1. 快點上學去，否則你就無法及時趕上第一堂課。
2. 要注意健康，這樣你才會過得快樂些。
3. 再過幾天，這家公司將會瀕臨破產的邊緣。

寫作參考

1. Go to school quickly, or you'll not be in time for the first class.
2. Take care of your health, and you'll lead a happier life.
   = Watch your health, and you'll lead a happier life.
   = Be careful of your health, and you'll lead a happier life.
3. A few more days, the company will be on the verge of bankruptcy.

▲ 活用句型

(1) Not... But... 並非……而是……

Example:

她不是我姊妹，而是我侄女。

She is not my sister but my niece.

(2) Not only... but (also)... 不僅……而且……

Example：

這小說家不僅在台灣有名，在歐洲也滿有名的。

The novelist is famous not only in Taiwan but (also) in Europe.

(3) Either... or... 要不……就是……

Example：

你能說法文或德語嗎？

Can you speak either French or Germany?

(4) Neither... nor... 既非……也非……

Example：

她和我都不了解他。

Neither she nor I can understand him.

★ 加強補充：以上四個活用句型，均為對等連接詞片語，可連接對等之單字、片語，或子句。

(1) 連接單字

Example：

他並非有錢而是窮。

He is not rich but poor.

均為形容詞

Example：

他不僅是個音樂家，而且是個商業人士。

He is not only a musician but (also) a business man.

均為名詞

Example：

你能以泰國文化唱歌或跳舞嗎？

Can you either sing or dance with Thailand culture?

均為動詞

Example：

你和我都沒錯。

Neither you nor I am wrong.

均為名詞

(2)連接片語

Example：

他不在臺灣，而在美國。

He is not in Taiwan but in U.S.A.

均為片語

Example：

他不僅以他的才能聞名，而且他的仁慈也衆所皆知。

He is famous not only for his talent but (also) for his kindness.

均為介詞片語

(3)連接子句

Example：

我想買這東西不是因爲它便宜，而是因爲它有用。

I want to buy this not because it is cheap, but because it is useful.

★ 加強補充：以上四個對等連接詞，其動詞依最近之主詞做變化。

Example：

不僅僅是你，我也該被譴責。

Not only you but I am to be blamed.

Example：

不是我，而是他該負責。

Not I but he is responsible for it.

Example：

要不是你，就是他已犯了錯。

Either you or he has made the mistake.

Example：

既非你，也非我有空。

Neither you nor I am free.

▲ 活用語法：... not only ... but (also) = not only ... but ... as well

Example：

他不僅聰明，而且仁慈。

He is not only clever but (also) kind

= He is not only clever but kind as well.

▲ not only...but (also)...之句型，若not only放在句首，其後要倒裝。換句話說，否定副詞放句首，其後採用倒裝句語法。

A. 遇be動詞時

Example：

他不僅勤勞，而且忠誠。

Not only is he industrious, but he is also faithful.

be動詞

☆ 請注意

but also若要連接主要子句，also要省略或置於句中。換言之，不可在but also後，直接接上主要子句。

Example：

Not only is he industrious, but he is also faithful.

Not only is he industrious, but he is faithful.

also省略

= Not only is he industrious, but he is faithful as well.

• 易犯錯誤範例

Example：

Not only is he industrious, but also he is faithful. (×)

→如果but also 之also沒有省略，不可在but also後，直接接上主要子句。

正 解：

Not only is he industrious, but he is also faithful. (○)

B. 遇助動詞時

Example:

她不僅會唱歌，她還會跳舞。

Not only can she sing, but she can dance as well.

助動詞

C. 遇一般動詞時

Example:

她不僅會唱歌，她還善舞。

Not only does she sing well, but she also dances beautifully.

寫作練習

1. 它並非詩人，而是小說家，這樣的說法並不過份。
2. Mary的工作不僅要處理難題，而且還要應付突來的挑戰。
3. 在我看來，你和他都沒有達到標準。
4. 他出名既不是因為他有學問，也不是因為他有錢。
5. 他不僅拒絕幫助我，而且還罵我。
6. 他失敗，不只因為他懶，而且是因為他不合群。

寫作參考

1. It is not too much to say that he is not a poet but a novelist.
2. Mary's job is not only to deal with problems but to meet the

unexpected challenges.

3. In my opinion, neither you nor he has met the standard.
4. He is famous neither because he is knowledgeable nor because he is rich.
5. Not only did he refuse to help me, but he scolded me.
6. He failed not only because he was lazy but because he was not sociable.

▶ Both...and... 二者皆

Both...and... 可連接對等之單字，片語、與子句。若連接主詞時，動詞一定爲複數。

1. 連接單字

Example:

他的演講既有趣又具教育性。

His lecture was both interesting and instructive.

Example:

他和我對於這結果都感滿足。

Both he and I are satisfied with the result.

2. 連接片語

Example:

這小說在情節與風格上，是非常優秀的。

The novel is very excellent both in plot and in style

3. 連接子句

Example：

他的成功是因勤勉和他有很多朋友幫忙他。

He successes both because he was industrious and because he had many friends to help him.

- ...as well as... ……和……一樣。
- ...as well as... 亦爲對等連接詞，與both...and相同，可連接對等之單字、片語、子句。
- ...as well as... 連接主詞時，動詞隨第一個主詞做變化。

Example：

他能說流利之俄文與英文。

He can speak good Russian as well as English.

- 易犯錯誤範例

Example：

He as well as I am satisfied with the result. (×)

動詞應搭配前者

正 解：

He as well as I is satisfied with the result. (○)

以第一個主詞為依據

寫作練習

1. 雖然我外向，（但）我也愛爵士樂和古典樂。
2. 他是知名的畫家及政治家。
3. 他這篇有關交通問題的論文，既有內涵，也令人印象深刻。
4. 想要學好翻譯，興趣及努力是你所需的。
5. 工人及學生全部參加了這次的民主運動。
6. 顯然他對音樂和繪畫都有興趣。

寫作參考

1. Although I am extrovert, I like both jazz and classical music.
2. He is known both as a painter and as a statesman.
3. His paper regarding traffic problem is informative as well as impressive.
4. Both interest and hard work are what you need to master translations.
5. Workers as well as students participated in the pro-democracy movements.
6. It is obvious that he is interested in music as well as (in) painting.
   = Obviously, he is interested in music as well as (in) painting.

▶ As soon as... 一……就……【以下列出多個通用句型】

Example:

他一來，Mary就離開。

As soon as The moment The instant The minute Once	he came, Mary left.

▲ As soon as為副詞連接詞，引導副詞子句，藉以修飾主要子句。

此句型亦可被以下連接詞所取代，均表示【一……就……】之含意。

No sooner...than...

= Hardly...when... 或Hardly...before...

= Scarcely...when... 或Scarcely...before...

Example:

我一到，Mary就離開。

No sooner had I come than Mary left.

否定副詞放句首，其後倒裝。

Example:

我一見到他，（我）就昏倒了。

No sooner had I seen him than I passed out.

= Hardly had I seen him when I passed out.

= Scarcely had I seen him before I passed out.

寫作練習

1. 當Tom一看到他父親，就揮手。
2. 當經理一聽到這悲傷的消息，就痛哭失聲。
3. 我一有空，就會打電話給你。
4. 我一轉身，就見到那搶匪一手拿著刀，向我衝過來。
5. 我一抵達New York，就見到一位學生，主動要當我的嚮導。

寫作參考

1. The moment Tom saw his father, he waved his hands.
2. Hardly had the manager heard the sad news when he burst out crying.
3. As soon as I have time, I will give you a buzz.
4. The instant I turned around, I saw a robber rushing toward me with a knife in his hand.
5. Scarcely had I arrived in New York before I saw a student who offered to be my guide.

▲ Not that… 並非…

Example:

並非我不喜歡那工作，而是我不夠資格。

Not that I dislike the task, but that I am not qualified for it.

▲ Now that⋯ 既然⋯

Example：

既然我有空，就可以享受音樂一下。

Now that I am free, I can enjoy music for a while.

= Since I am free, I can enjoy music for a while.

= Seeing that I am free, I can enjoy music for a while.

寫作練習

1. 既然颱風就要來了，你為什麼還要去遠足呢？
2. 信不信由你，並非我不愛Mary，而是我更愛我的國家。
3. 既然你已是個大學生，就應當學著獨立，不靠父母的幫忙。
4. 你不該因為你比他聰明就驕傲。
5. 我不會因為他窮，就輕視他。

寫作參考

1. Now that typhoon is coming, why is it that you still want to go hiking?
2. Believe it or not, not that I don't love Mary, but that I love my country more.
3. Now that you are a college student, you should learn to be independent of your parents' help.

4. You should not be proud because you are cleverer than he.

5. I won't look down upon him because he is poor.

▶ so (in order) that...may 以便於……
Lest...should 以免……萬一……

Example：

詳細閱讀問題，以期在考試時你較不會犯錯。

Read the questions carefully so that you may not make mistake in the test.

Example：

我母親努力工作，以期能支撐我們。

My mother works hard in order that she may support us.

Example：

把那照片藏好，以免他萬一見到。

Hide the picture lest he should see it.

▲ so that..may 或in order that所引導之副詞子句中，若主詞與主要子句相同，則可變成不定詞片語。換句話說，可使用以下句型：

so as to + 原型動詞 = in order to + 原型動詞

Example：

開那輛全新車子要小心點，以免犯任何錯。

Drive the brand new car with care so as not to make any mistakes.

= Drive the brand new car carefully in order not to make mistake.

Example：

我來此，爲的是希望能見到Mary。

I came here in order that I might see Mary.

= I came here in order to see Mary.

特殊句型

Example：

我去那兒，爲的是見到Mary。

I went there so as to see Mary.

= I went there with a view to seeing Mary

= I went there with an eye to seeing Mary.

▶ lest...should 以免……萬一

lest爲副詞連接詞，引導副詞子句，在該子句中，只能用助動詞should。但是，should亦可予以省略，而直接用原形動詞。

Example：

我早起，以免趕不上火車。

I got up early lest I (should) miss the train.

Example：

他早起，以免萬一遲到。

He rose early lest he (should) be late.

▲ 當使用lest...should句型時，可以被下列句型取代。

Example：

他設法完成工作，以免萬一被處罰。

He managed to finish the work on time lest he should be punished.

= He managed to finish the work on time for fear that he might be punished.

= He managed to finish the work on time for fear of being punished.

寫作練習

1. 我會協助你的工作，以便我們能早點回家。
2. 我們說話要小聲點，以免把嬰兒吵醒。
3. 請事先把稿子檢查一遍，以免有錯。
4. 我決定學攝影，以便更能欣賞自然之美。

寫作參考

1. I will help you with your work so that we may go home early.
2. We talk in a low voice lest we should wake the baby up.

   或

   We talk in a low voice lest we should awake the baby.
3. Please check the manuscript in advance lest there should be errors.
4. I have decided to learn photography in order that I may better appreciate the beauty of nature.

▶ So...that... 如此……以致於

常和Such...that... 交替使用，兩者意思相同。

Example:

這隻狗如此溫和，以致於我不怕它。

The dog is so gentle that I am not afraid it.

Example:

她是位如此好的鋼琴演奏者，以致於她所有的朋友都希望聽她演奏。

She is such a good piano player that all her friends want to listen to her play.

▲ So爲副詞，其後接形容詞或副詞，來強化其意涵。

Example:

Jack is so nice that everybody likes him.

↖ 接形容詞

Example:

He works so hard that he passed the exam.

↖ 接副詞

☆ **請注意**

(1) Such...that... 與so...that...句構完全一樣，只是such爲形容詞，其後只能接名詞，而that爲副詞連接詞，引導副詞子句，修飾such。

(2) Such ＋ 名詞 ＋ that⋯

Example:

He is such a nice boy that we all like him.

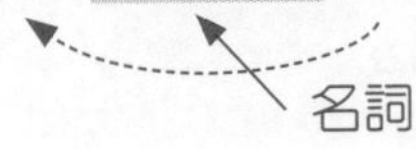

(3) 在so...that... 或 such...that... 句構中，若that子句中之主詞與主要子句相同時，that子句可簡化爲as to + 原型動詞之型式。

Example：

他是如此勇敢，以致於贏得我的尊重。

He is so brave that he wins my respect.

= He is so brave that as to win my respect.

Example：

他是一位好學生，以致於他被指定爲班代表。

He is such a good student that he is assigned to be the class leader.

= He is such a good leader as to be the assigned class leader.

(4) ..., so that... = ..., so... 因此……（so that之前若有逗點，則等於so表示因此……）

Example：

他發音清楚，因此每一位都能聽到。

He pronounced clearly, so that everybody could hear him.

= He pronounced clearly, so everybody could hear him.

(5) ...so...that句型中，原則上，so之後原本只可接形容詞或副詞，但亦可接單數可數名詞。換言之，含有不定冠詞a之名詞。

◎ 實用句型

...so + 形容詞 + a + 名詞 + that...

- 易犯錯誤範例

Example: 1

他是位如此好的男孩，以致於我們喜歡他。

He is a so good boy that we like him. (×)

正 解:

He is so good a boy that we like him. (○)

↖ 形容詞 + a + 名詞 + that ...

Example: 2

他們是如此好的男孩們，以致於我們喜歡他們。

They are so good boys that we like them.

↖ 不可在此使用複數

正 解:

They are such good boys that we like them.

Example: 3

這是如此好的音樂，以致於我喜歡他。

It is so good music that I enjoy it. (×)

↖ 僅可在此使用單數可數名詞

正 解:

It is such good music that I enjoy it. (○)

(6)換言之，such之後可接任何名詞（可數或不可數）。

Example:

He is such a good boy that we like him. → 單數可數名詞

They are such good boys that we like them. → 複數可數名詞

It is such good music that I enjoy it. → 不可數名詞

(7)so可修飾表數量之形容詞，例如：many、much、few、little。

→...so many + 複數名詞 + that... 如此多的……以致於

→...so much + 不可數名詞 + that... 如此多的……以致於

→...so few + 複數名詞 + that... 如此少的……以致於

→...so little + 不可數名詞 + that... 如此少的……以致於

Example:

我有如此多的工作，以致於我無法和你一起去。

I have so much work to do that I can't go with you.

= I have so many things to do that I can't go with you.

Example:

時間所剩無幾，以致於我不認為我能如期完成。

There is so little time left that I don't think I can finish it as scheduled.

Example:

他只有極少數的朋友，以致於他的生活孤獨。

He has few friends that his life is lonely.

▲ so或such引導之詞類，置於句首時，採倒裝寫法。

A. Be動詞

Example：

他是如此友善，以致於我喜歡他。

He is so nice that I like him.

So nice is he that I like him.

Example：

他是一位如此友善的男孩，以致於我喜歡他。

He is such a nice boy that I like him.

Such a nice boy is he that I like him.

B. 助動詞

Example：

他能說如此多國語言，以致於他環遊世界應該沒有問題。

He can speak so many languages that he should have no problem travelling around the world.

So many language can he speak that he should have no problem travelling

採倒裝寫法

around the world.

C. 一般動詞倒裝時，要與do、does、did並用。

Example：

他歌唱得如此好，以致於他能變成一位歌手。

He sings so well that he can become a good singer.

So well does he sing that he can become a good singer.

Example：

他寫出這麼好的文章，以致於他贏得獎項。

He wrote such a good article that he won the award.

Such a good article did he write that he won the award.

寫作練習

1. Tom如此深愛著他的女朋友，以致於他決定在情人節，向她求婚。
2. 這是令人興奮的比賽，以致於我把即將來臨之期末考，忘得一乾二淨。
3. 他如此言而有信，因此你可以把任務交付給他。
4. 他們對這消息感到如此欣喜，以致於睡不著覺。
5. 這是個大好機會，因此我們不能錯過。
6. 他這麼懶，以致於大家都輕視他。

寫作參考

1. Tom loves his girlfriend so much that he decides to propose to her on Valentine's day.
2. So exciting was the game that I forgot all about the coming finals.

3. He is so trustworthy that you can trust him with the mission.
4. They were so glad at the news that they couldn't sleep.
5. So good is the chance that we can't miss it.
6. So lazy is he that everyone looks down upon him.

▲ As 取代though之用法

Though引導之副詞子句，若有下列句構，可用as取代。

A. Though + 主詞 + be動詞 + 形容詞 = 形容詞 + as + 主詞 + be動詞

Example:

雖然他窮，他不覺得（是）次等。

Though he is poor, he doesn't feel inferior.

= Poor as he is, he doesn't feel inferior.

Example:

雖然他疲憊，他仍持續做那件工作。

Though he was tired, he kept doing it.

= Tired as he was, he kept doing it.

B. Though + 主詞動詞 + be 動詞 + 名詞

= 名詞 + as + S + be 動詞

Example:

雖然他們是好學生，他們有時仍會犯錯。

Though they are good students, they sometimes make mistakes.

= Good students as they are, they sometimes make mistakes.

→ 以上用法，be動詞後之名詞若為單數，移至句首後，冠詞要省略。

• **易犯錯誤範例**

Example:

雖然他是好學生，他有時仍會犯錯。

A good student as he is, he sometimes makes mistakes. (×)

↖ 冠詞要省略

正 解:

Though he is a good student, he sometimes makes mistakes. (○)

= Good student as he is, he sometimes makes mistakes. (○)

C. Though + 主詞 + 動詞 + 副詞 = 副詞 + as + 主詞 + 動詞

換句話說，在though之句構中，有副詞時，要將副詞移至句首，再將though改為as即可。

Example:

雖然他認眞工作，（但）他考試失敗了。

Though he worked hard, he failed in the exam.

↖ 副詞

= Hard as he worked, he failed in the exam.

寫作練習

1. 雖然他很嚴格，但富有同情心。
2. 雖然他是懦夫，但無法忍受這樣的侮辱。
3. 雖然他的手稿很珍貴，（但）他還是把它燒掉了。
4. 雖然他的功課很好，卻一點都不傲慢。

寫作參考

1. Though he is stern, he is full of sympathy.

   = Stern as he is, he is full of sympathy.
2. Coward as he is, he can't bear such an insult.
3. Precious as his manuscript was, he put it in the fire.
4. Though he does well in study, he is not arrogant.

   = Well as he does in study, he is not arrogant.

Module

02

Business Writing Essentials
基本商業書信

Module 02

Chapter ⑪ Recommendation Letter 1（推薦函1）

Chapter ⑫ Recommendation Letter 2（推薦函2）

Chapter ⑬ Application Letter（申請函）

Chapter ⑭ Job Interview Letter（工作面試函）

Chapter ⑮ Re-Scheduling Letter（重新預約函）

Chapter ⑯ Sample Request Letter（樣品需求函）

Chapter ⑰ Relocation Letter（搬遷函）

Chapter ⑱ Sick Leave Letter（病假函）

Chapter ⑲ Experienced French Instructor Wanted Letter（徵聘有經驗之法文教師）

Chapter ⑳ Signing MoU（簽署備忘錄）

Chapter 1 Greeting Letter 1（歡迎信1）

Scenario（情境）：

第一次與對方聯繫，並打算前往拜會。

Dear President Sieng Sovanna, October 29, 2016

Greeting from Taiwan!

This is Dean Eric Chu of National Taichung University of Education (NTCU), Taiwan. I am so glad that we are going to meet around January 16th, 2017.

In addition, I extremely anticipate to take this opportunity paying a visit to the Ministry of Education of your country for the future bilateral cooperation. NTCU is a 116 years old university and we are more than happy to share what we had in the past 10 decades among the global villages. Please kindly also transfer this message to the officers at the Ministry of Education.

I visited the National University of Laos, Vice-Minister of Ministry of Education of Laos, and Minister of Technology of Laos around August, 2011. Currently, we had a very strong relationship with President Prof. Dr. Soukkongseng SAIGNALEUTH of National University of Laos during the last visit.

Your kind arrangement to visit the Ministry of Education will be an extraordinary benefit for both countries for cultural exchanges in the near future.

Best regards,

Eric Chu

Dean Eric Chu

National Taichung University of Education (NTCU), Taiwan

Vocabulary or Phrases:

1. extremely 非常地
2. paying a visit to 訪問、參訪
3. Best regards 致上我最高的問候

主旨

來自台灣的問候！

我是台灣國立台中教育大學（NTCU）的朱海成處長。我很高興我們將會在2017年1月16日會面。

此外，我非常期待藉這個機會，為了未來雙邊的合作，能夠拜訪貴國的教育部。NTCU是一所具有116年歷史的大學，我們非常樂意分享過去十年來，我們在地球村的貢獻與成就。請您也將此信息轉達給教育部官員。

我在2011年8月左右拜訪過寮國國立大學、寮國教育部副部長，以及寮國的技術部部長。自從上次的拜訪，我們與寮國國立大學的Soukkongseng SAIGNALEUTH校長有著非常友好的關係。

您安排的教育部參訪，將在不久的將來，為兩國的文化交流帶來很大的益處。

致上我最高的問候，

Chapter 2 Greeting Letter 2（歡迎信2）

Scenario（情境）：

國內學術單位第一次與國外學術單位聯繫，希望與柬埔寨國立大學建立第一次接觸之歡迎信。

Dear President SiengSovanna, October 29, 2016

Greeting from Taiwan!

This is the Dean of International Affairs, Dr. Eric Chu, of National Taichung University of Education, hereinafter called NTCU, Taiwan. The visiting mission is planning to officially visit you around January 16th, 2016.

In addition, I extremely anticipate taking this opportunity to pay a visit to the Ministry of Education of your esteemed country for the future bilateral academic cooperation. Phenomenally, NTCU is a 115 years old university with glorious history and we are more than happy to share what we had achieved over the past 10 decades among the global villages. Please kindly transfer this message to the officers at the Ministry of Education.

I visited the National University of Laos, Vice-Minister of Ministry of

Education of Laos, and Minister of Technology of Laos around August, 2011. Currently, we had a very strong relationship with President Prof. Dr. Soukkongseng SAIGNALEUTH of National University of Laos during the last visit.

Your kind arrangement to visit the Ministry of Education will be an extraordinary benefit for both countries for cultural exchanges in the near future.

Best regards,

Eric Chu

Dean Eric Chu

National Taichung University of Education (NTCU), Taiwan

140 MinSheng Road, Taichung, Taiwan, 40306

Dean Office direct line: (04)2222-2222 or (04)2222-3333 ext. 11

Vocabulary or Phrases:

1. hereinafter 以下、在下文中
2. visiting mission 訪問使命
3. in addition / additionally 此外
4. phenomenally / remarkably / extraordinarily / outstandingly / unbelievably / astonishingly 驚人地
5. extraordinary / surprising 非凡地

主旨

來自台灣的問候！

我是在台灣國立台中教育大學（以下簡稱NTCU）擔任國際事務處處長的朱海成博士。本校參訪團計畫於2013年1月16日左右前去正式拜訪您。

此外，我非常希望藉這個機會，為了未來雙邊的學術合作，能夠拜訪貴國的教育部。令人印象深刻地，NTCU是一所具有115年光榮歷史的大學，我們很樂意分享過去十年來，我們在地球村的貢獻與成就。請您將此信息轉達給教育部官員。

我在2011年8月左右拜訪過寮國國立大學、寮國教育部副部長，以及寮國的技術部部長。自從上次的拜訪，我們與寮國國立大學的Soukkongseng SAIGNALEUTH校長有著非常友好的關係。

您安排的教育部參訪，將在不久的將來，為兩國的文化交流帶來很大的益處。

致上我最高的問候，

Chapter 3 Invitation Letter（邀請函）

Scenario（情境）：

邀請美國設計公司到公司生產線參觀

To: CEO Mrs. Becky Herman

From: Eric Chu

Date: July 2, 2015

Subject: Quotation Inaccuracy

Dear Mrs. Herman,

This is marketing manager, Mr. Eric Chu, who is also a project manager of Twin Shoes Inc., China. We have solid assembly lines as well as flexible manufacturing system to reduce the time to market. As a leading OEM company in South East China, we would like to take this precious opportunity to sincerely invite you paying a visit to our factories in Shenzhen, China.

Undoubtedly, in order to express our sincerity and hospitality, we would like to offer you the round-trip business class airplane tickets with 5 stars Hotel during your stay.

Please kindly inform me when your traveling itinerary is settled down. Furthermore, please do not hesitate to contact with me if there is any inquiry arises. At last, we wish that we can see you soon here.

Cordially,

Eric Chu

Eric Chu

Marketing Manager

Black Technical Enterprises

P.O. Box 123 • Keeseville • NY • 12944

Phone 518-555-2345

主旨

親愛的Herman女士：

我是行銷部經理朱海成先生，同時也是中國Twin Shoes公司的專案經理。我們有可靠的組裝生產線，以及富有彈性的製造系統，用以減少產品上架時間。作為在中國東南最主要的OEM（原始設備製造商）公司，我們想利用這個寶貴的機會，誠摯地邀請您前來參訪我們在中國深圳的工廠。

毫無疑問地，為了表示我們的誠意和熱情，我們願意為您提供往返的商務艙機票及五星級飯店的住宿。
當您的旅行行程確定後，請通知我。此外，如果有遇到任何問題，請不要猶豫，直接與我聯繫。最後，希望我們能很快與您相見。

誠摯地歡迎您，

Chapter 4 Curriculum Vitae（自我介紹）

Scenario（情境）：
個人學經歷自我介紹

Dean HaiCheng Eric Chu

Professor at National Taichung University of Education

Dean of International Affairs
Dean of Research & Development
140 Min-Shen Road, Taichung, Taiwan, 40306
e-mail: ayura66@gmail.com
Cellular: +886-911-161-545

Dean HaiCheng Eric Chu received his Ph.D. degree in System Science and Industrial Engineering in 1996 and Master of Computer Science in 1992 from SUNY at Binghamton respectively. He was a senior software engineer at Cheyenne Software Inc., New York, USA in 1996. Dr. Chu was lecturing at University of Northern British Columbia (UNBC) in Canada in 2008. Currently, Dean Chu is a full professor at the Department of International Business of NTCU, Taiwan. Dean Chu attended Harvard Business School for PCMPCL V program in 2007. He has authored several textbooks respecting Management Information System, e-Commerce, Global Logistics Management, Commercial Automation, System Analysis and Design.

1. PERSONAL

Birth: 09th April , 1966, Taipei, Taiwan

Citizenship: Taiwan

2. EDUCATION

Doctor Degree from State University of New York at Binghamton, USA – System Science and Industrial Engineering. (01/1993 ~ 01/1996)

Master Degree from State University of New York at Binghamton, USA – Computer Science. (09/1991 ~ 12/1992)

Bachelor Degree from Tunghai University, Information Science, Taiwan. (09/1985 ~ 06/1989)

3. EMPLOYMENT

(a) *Academic appointments*

1. National Taichung University of Education (NTCU), Taiwan
   Full Professor, Department of International Business (2012 till now).
2. Dean of International Affairs / Dean of Research & Development (2012 till now).
3. Tunghai University, Taiwan
   Associate Professor, Department of International Trade, (2002- 2010).

4. FengChia University, Taiwan

Associate Professor, Department of International Trade, (2002- 1999).

(b) *Business appointments*

Senior Software Engineer at Cheyenne Software Inc., New York, USA (was a NASDAQ Inc.) merged by Computer Associate Inc. (05/1995 ~ 07/1996). Dr. Chu is the leading MIS/e-Commerce consultant in Taiwan business corporations.

4. SCHOLARLY ACTIVITIES

(a) *Areas of expertise*

Harvard Business School - Fifth class of Program on Case Method and Participant-Centered Learning (PCMPCL), July 29th ~August 9th, 2007, Boston, USA.

(b) *Research grants*

National Science Council (Taiwan)

National Science Foundation (USA)

5. References

• Dr. Edward Guiliano

President

Northern Boulevard

P.O. Box 8000

Old Westbury, NY 11568-8000

Tel 516.688.7777

Fax 516.688.9999

edwardg0967@nyit.edu

• Dr. Rahmat Shouresh

Provost and vice president for academic affairs

Northern Boulevard

P.O. Box 8000

Old Westbury, NY 11568-8000

Tel 516.686.6666

Fax 516.686.2345

rshoures0976@nyit.edu

• Dr. Shawn Chen

President

SIAS International University

168 Renmin Road East, Xinzheng

Zhengzhou City, Henan, China

Tel 0371-62648888

shawn_sias@yahoo.com

Vocabulary or Phrases:

1. respectively 分別地
2. respecting 有關於

主旨

朱海成處長從美國紐約州立大學Binghamton分校畢業，分別在1992年取得電腦科學碩士、1996年取得系統科學與工業工程博士學位。1996年他在美國紐約Cheyenne軟體股份有限公司，擔任高級軟體工程師。朱博士2008年在加拿大UNBC講課。目前，朱處長為NTCU國際企業學系的正教授。朱院長2007年參加了哈佛商學院PCMPCL V計畫。他撰寫了許多有關管理資訊系統、電子商務、全球運籌管理、商業自動化、系統分析和設計等教科書。

1. 個人資訊

出生：1966年04月09日，台北，台灣

國籍：台灣

2. 學歷

美國紐約州立大學賓漢姆頓分校，系統科學與工業工程博士學位。（1993/01～1996/01）

美國紐約州立大學賓漢姆頓分校，計算機科學碩士學位。（1991/09～1992/12）

台灣東海大學，資訊工程學系學士學位。（1985/09～1989/06）

3. 資歷

(a) 學術任職

1. 台灣國立台中教育大學（NTCU），國際企業學系教授（2012年至今）。
2. 國際事務處處長（2012年至今）。
3. 台灣東海大學
   副教授，國際貿易系（2002-2010）。
4. 台灣逢甲大學
   副教授，國際貿易系（2002-1999年）。

(b) 企業任職

美國紐約Cheyenne軟體公司（為納斯達克公司，後被美國國際聯合電腦公司併購），高級軟體工程師（05/1995～07/1996）。朱博士在台灣的工商企業，為卓越的MIS/電子商務顧問。

4. 學術活動

(a) 專業知識方面

美國波士頓哈佛商學院（PCMPCL），2007/7/29～8/9。

(b) 研究補助金

國家科學委員會（台灣）

美國國家科學基金會（美國）

Chapter 5 Job Offer Letter 1（錄用函1）

Scenario（情境）：
公司通知應徵者至公司報到

Black Technical Enterprises

P.O. Box 123 • Keeseville • NY • 12944

Phone 518-555-2345

March 1, 2016

John Chen

513, Sec. 1, Taiwan Boulevard, Taichung City, 43301 Taiwan

Dear Mr. John Chen,

JOB OFFER

Black Technical Enterprises, Inc. is pleased to offer you a job as a Senior Software Engineer. We trust that your knowledge, skills and experience will be among our most valuable assets.

Should you accept this job offer, per company policy you'll be eligible to receive the following beginning on your hire date.

- **Salary**: Annual gross starting salary of $83,000, paid in biweekly installments by your choice of check or direct deposit
- **Performance Bonuses**: Up to three percent of your annual gross salary, paid quarterly by your choice of check or direct deposit

- **Benefits**: Standard, **Black Technical Enterprises**-provided benefits for salaried-exempt employees, including the following
  - o Child daycare assistance
  - o Education assistance
  - o Health insurance
  - o Sick leave
  - o Two weeks paid vacation and personal days

To accept this job offer:

1. Sign and date this job offer letter where indicated below.
2. Sign and date the enclosed Non-Compete Agreement where indicated.
3. Sign and date the enclosed Confidentiality Agreement where indicated.
4. Sign and date the enclosed At-Will Employment Confirmation where indicated.
5. Mail all pages of the signed and dated documents listed above back to us in the enclosed business-reply envelope, to arrive by Friday, November28, 2016. A copy of each document is enclosed for your records.
6. Attend new-hire orientation on Tuesday, December 1, 2016, beginning at 8:00 AM sharp.

To decline this job offer:

1. Sign and date this job offer letter where indicated below.

2. Mail **all pages** of this job offer letter back to us in the enclosed business-reply envelope, to arrive by Thursday, November 28, 2016.

If you accept this job offer, your hire date will be on the day that you attend new-hire orientation. Plan to work for the remainder of the business day after new-hire orientation ends. Please read the enclosed new-hire package for complete, new-hire instructions and more information about the benefits that **Black Technical Enterprises** offers.

We at **Black Technical Enterprises** hope that you'll accept this job offer and look forward to welcoming you aboard. Your immediate supervisor will be Jennifer Kermer, Manager, the Department of Engineering. Please feel free to call Jennifer or me if you have questions or concerns. Call the main number in the letterhead above during normal business hours and ask to speak to either of us.

Sincerely,

Kevin Bruce

Kevin Bruce

Hiring Coordinator, Human Resources

Enclosures: 8

Accept Job Offer

By signing and dating this letter below, I, John Chen, accept this job offer of Senior Engineer by **Black Technical Enterprises**.

Signature: ____________________ Date: ____________________

Decline Job Offer

By signing and dating this letter below, I, John Chen, decline this job offer of Senior Engineer by **Black Technical Enterprises**.

Signature: ____________________ Date: ____________________

Vocabulary or Phrases:

1. assets 資產
2. installments 分期付款
3. direct deposit 直接存款
4. quarterly 每季地
5. Sick leave 病假
6. personal days 事假
7. Non-Compete Agreement 競業條款
8. Confidentiality Agreement 保密協議
9. Orientation（新生、新會員的）輔導、（對新環境的）適應
10. Decline 婉拒、下降
11. look forward to 期待
12. immediate supervisor 上級主管
13. letterhead 信箋抬頭

主旨

尊敬的John Chen先生：

工作機會

Black科技企業公司很高興為您提供一份工作。作為一名高級軟體工程師，我們相信以您的知識、技術和經驗，將能成為我們最寶貴的資產之一。

如果您接受這個工作機會，依照每家公司的政策，您在就職後即有資格獲得以下福利。

- 薪酬：年起薪$83,000，每雙週支付一次，由您選擇支票或直接存款。
- 績效獎金：高達年工資總額的3%，按季支付，由您選擇支票或直接存款。
- 福利：基本上，Black科技企業公司提供員工的福利包含以下幾點。
  - o 兒童日間照護援助
  - o 教育援助
  - o 健康保險
  - o 病假
  - o 兩個星期的有薪休假和事假

若要接受這個工作機會：

1. 於此錄取通知書下方處簽名並註明日期。
2. 於隨函附上的競業限制合約上簽名並註明日期。
3. 於隨函附上的保密協定上簽名並註明日期。

4. 於隨意雇傭確認書上簽名並註明日期。
5. 將以下已簽名並註明日期的所有文件置於隨函覆上的商業回復信中，且在2016年11月28日前寄達。請隨函附上每一份文件的副本以便作為你的紀錄。
6. 出席2016年12月1日星期二的新員工就職輔導，活動於上午8:00整開始。

若要拒絕這個工作機會：
1. 於此錄取通知書下方處簽名並註明日期。
2. 將此郵件內的所有文件置於隨函覆上的商業回復信中，並於2016年11月28日星期四前寄達。

如果你接受這個工作機會，你的就職日期將為你參加新員工就職輔導的那天。為了迎接新員工就職培訓結束後剩餘的工作天，請閱讀所附的新員工資訊，以了解完整的相關指示說明及更多Black科技企業提供的福利信息。

我們Black科技企業希望您能接受這個工作邀請，並期待著您的加入。你的上級主管為工程部門經理，Jannifer Kermer。如果您有問題或疑慮請隨時致電Jannifer，或請於正常上班時間撥打於信箋抬頭所附的電話，並與我們兩人之一聯繫。

誠摯的，

接受工作邀請

藉由在這封信下方簽名並註明日期，我，Mr. John Chen，接受Black技術企業的高級工程師一職。

簽名：＿＿＿＿＿＿＿＿＿＿＿＿日期：＿＿＿＿＿＿＿＿＿＿＿＿

拒絕工作邀請

藉由在這封信下面簽名並註明日期，我，Mr. John Chen，拒絕Black技術企業的高級工程師一職。

簽名：＿＿＿＿＿＿＿＿＿＿＿＿日期：＿＿＿＿＿＿＿＿＿＿＿＿

Chapter 6 Job Offer Letter 2（錄用函2）

Scenario（情境）：
公司通知應徵者至公司報到

iDigital Technical Enterprise

No. 144, MinSheng Road, Taichung City, Taiwan, ROC

e-mail: hr@idigital.com.tw

Tel: +886-4-2218-1122

Fax: +886-4-2218-1123

Joyce Smith

Northern Boulevard

P.O. Box 6000

Old Westbury, NY 11568-6000, USA

Dear Mrs. Smith:

I am pleased to send this notification on behalf of our company to you regarding the offer of a permanent position with **iDigital Technical Enterprise** as a Public Relationship Manager starting on January 1st, 2017.

Please kindly find the following statements that we fully discussed during the interview:

Compensation:

Your initial compensation will be USD$ 4,000 / month. The probationary period is three months as we mentioned during the interview. Other benefits will be enforced momentarily once you succeed in fulfilling the job descriptions after the probation period.

Benefits:

Demonstrably, we cover your health and dental insurances. For those employees, who complete the probation period will be entitled to have 2 weeks paid vacation, and contributions to a retirement fund.

As aforementioned statements, once you pass the probation period, your compensation package will thereafter be reassessed on an annual basis.

We are looking forward to working with you in this promising and energetic working team.

Sincerely,

Becky Chen

Becky Chen

Manager of Human Resources

I sincerely invite you to peruse the contents of this employment offer letter. To accept, please sign and date the letter below with scanning it back to **iDigital Technical Enterprise** before December 28th.

I, Joyce Smith, agree to accept the position offered by **iDigital Technical Enterprise** and the aforementioned terms and policies indicated above.

(Signature)   (Date)

P.S.: If you would like to decline this offer letter, please kindly reply me with a brief letter.

Vocabulary or Phrases:

1. on behalf of 代表
2. probationary period 試用期
3. momentarily 立刻
4. demonstrably 可證實的
5. retirement fund 退休金
6. aforementioned 前述的
7. thereafter 之後、隨後
8. reassessed 重新評估
9. promising 有前途、有希望的
10. energetic 有活力的、精力充沛的
11. peruse 細讀

主旨

親愛的Smith女士：

我很高興代表本公司發送此郵件通知您從2017年1月1日起，你將成為我們公司永久的公共關係經理。

請找到以下我們在面試過程中充分討論過的聲明：

酬金：

您最初的薪資將會是4,000美元一個月。試用期如面試時所說，為三個月。其他福利將會在您確實完成工作並通過試用期後立即生效。

福利：

可證實地，我們包含了您的醫療及牙醫險。對於那些通過試用期的員工，將有權獲得兩星期的有薪假以及退休金。

正如前述所言，一旦您通過試用期，您的薪資隨後將會每年重新評估。

我們很期待和您一起在這個充滿前途且活躍的公司工作。

我誠摯的邀請您細讀此封錄用函的內容。請在以下空格處簽名並寫上日期以表示同意，並於12月28日前以掃描的方式寄回給我們。

我同意接下在iDigital Technical Enterprise的職位，以及前述所提到的條件和政策。

Chapter 7 Rejection Letter 1（拒絕函1）

Scenario（情境）：

公司通知應徵者沒被錄用

To: Teresa May

From: Becky Chen

Date: December 10

Subject: Re: Application

Dear Ms. May,

Thank you for your response to our recent advertisement offering a management trainee position. Your credentials are impeccable, and I am sure you would be a credit to any company with which you are associated.

Although our current vacancy has been filled by another candidate, we will keep your application on file in case we have other opening in the near future.

Thanks again for your application, and good luck with your job search.

Becky Chen

Becky Chen

Vocabulary or Phrases:

1. response 回覆
2. credentials 憑證、資格
3. impeccable 無缺點的、無瑕疵的
4. vacancy 空缺
5. keep... on file 保存備查

主旨

親愛的May小姐：

感謝您回應我們最近管理培訓生的職位廣告。您的資格及憑證無可挑剔，而且我相信您對於任何與您接觸過的公司都是一項榮譽。

雖然我們目前的空缺已有另一名候選人補上，我們會將您的申請資料保存好，以防我們在不久的將來有其他的職缺。

再次感謝您的申請，祝福您尋找工作能夠順利。

Chapter 8 Rejection Letter 2（拒絕函2）

Scenario（情境）：
公司通知應徵者沒被錄用

To: Teresa May

From: Becky Chen

Date: August 21st

Subject: Please keep in touch

Dear Ms. May,

I am writing at this moment to inform the result of your interview at **iDigital Technical Enterprise** last week. We have made our decision to go with another applicant. Undoubtedly, based on your interview, we have concluded that you are a first-rate candidate.

Frankly speaking, you are the one of the most promising candidates and we really had a hard time to choose from the interviews. I would like to personally thank you for taking your precious time for the interview.

We wish you all the best in your job hunting. Please keep in touch for the next job opening.

Becky Chen

Becky Chen

Human Resource Manager

Vocabulary or Phrases:

1. first-rate 第一流的
2. Frankly speaking 坦白來說
3. promising 有前途的、有希望的

主旨

至：Teresa May

自：Becky Chen

日期：8月21日

主題：請保持聯繫

親愛的May小姐：

我在此時寫信來是為了告知您關於上週在iDigital Technical Enterprise的面試結果。我們已經決定錄用另一位申請者。毫無疑問地，基於您在面試的表現，我們將您視為一流的候選人。

坦白來說，您被我們視為最有前途的候選人之一，而這也使得我們在面試中花了很長的時間才做出決定，本人由衷地感謝您撥出寶貴的時間來參與此次的面試。

我們祝福您找工作順利。請保持聯繫，以利下一個職位空缺時。

Becky Chen

Becky Chen

人事部門經理

Chapter 9 Rejection Letter 3（拒絕函3）

Scenario（情境）：
公司通知應徵者沒被錄用

To: Teresa May

From: Becky Chen

Date: December 10

Subject: Re: Application

Dear Ms. May,

Thank you for your response to our recent advertisement offering a management trainee position. Your credentials are impeccable, and I am sure you would be a credit to any company with which you are associated.

Although our current vacancy has been filled by another candidate, we will keep your application on file in case we have other opening in the near future.

Thanks again for your application, and good luck with your job search.

Becky Chen

Becky Chen

Human Resource Manager

Vocabulary or Phrases：

1. response 回覆
2. credentials 憑證、資格
3. impeccable 無缺點的、無瑕疵的
4. vacancy 空缺
5. keep... on file 保存備查

主旨

親愛的May小姐：

感謝您回應我們最近所提供的培訓生管理這個職位的廣告。您的資格都無可挑剔的，而我相信您對於任何與您有關的公司來說都是項榮譽。

但由於我們目前的職缺已被另外一位候選人補上，我們會將您的申請書保存備查以利日後有其他的職缺開放時之用。

再次感謝您的申請，也祝您尋找工作順利。

Chapter 10 Fax Communication letter（傳真信函）

Scenario（情境）：

以傳真方式通知國外客戶

Black Technical Enterprises ***FAX***

To: Vice President Teresa May

Date: July 6th, 2016

Company: Global Sources Inc.

Phone: 516.688.7777

Fax: 516.688.9999

Total pages: 1

Sender: Black Technical Enterprises

Address:140 Min-Shen Road, Taichung, Taiwan, 40306

Phone: +886-4-2218-5566

Fax: +886-4-2218-4422

Dear Vice President Teresa,

Please kindly find the fax, which is the confirmation of your PO (Purchase Order).

We optimistically estimate the shipment will take two weeks.

Our representative will contact with your secretary momentarily.

Best,

Kevin Bruce

Kevin Bruce

Marketing Manager

Vocabulary or Phrases:

1. confirmation 確認
2. Purchase Order (PO) 訂單
3. optimistically 樂觀地
4. momentarily 立刻地、隨時

主旨

尊敬的Teresa副總裁：

敬請檢視傳真，這是您的PO（採購訂單）的確認。

我們樂觀地估計發貨需要兩週時間。

我們的代表將與您的秘書隨時聯繫。

祝好，

Kevin Bruce

Kevin Bruce

營銷經理

Chapter 11 Recommendation Letter 1（推薦函1）

Scenario（情境）：
推薦求職者之信件

To Whom It May Concern: Tuesday, July 07, 2015

I have been working closely with Steve Wang in the past two years at Black Technical Enterprises. Steve completed remarkable assignments in our R&D division since he joined us. He is second to none in the related fields. Regrettably, he is leaving this company.

Unquestionably, Steve has proven himself to be an outstanding staff in our company. His sophisticated knowledge of pharmacy is superior to many people in today's market. Furthermore, his working attitude is exceptional with relentless pursuit of perfection.

Consequently, I would like to recommend him without any reservation. If you have any questions, please do not hesitate to contact with me.

Yours faithfully,

Becky Chen

Director of R&D

Black Technical Enterprises

Vocabulary or Phrases:

1. Remarkable 傑出的
2. second to none 首屈一指的
3. Regrettably 很遺憾地
4. Unquestionably 毫無疑問地
5. proven himself 自我證明
6. outstanding 優秀的
7. sophisticated 成熟的、經驗豐富的
8. Furthermore 此外
9. Exceptional 卓越的、特殊的
10. relentless pursuit of perfection 對完美的不懈追求
11. Consequently 因此

主旨

敬要啓者：　　　　　　　　　　　　　　　　　週二，2015年7月7日

過去的兩年中，我一直與史蒂夫王在Black科技企業密切合作。自從史蒂夫加入了我們，他完成了我們的研發部門傑出的任務。他在相關領域是首屈一指的。遺憾的是，他將要離開這家公司。

毫無疑問，史蒂夫已經證明了自己是我們公司的優秀員工。他成熟的藥學知識優於相同領域的許多人。此外，他的工作態度是卓越與完美的不懈追求。

因此，我沒有任何保留地想推薦他。如果您有任何疑問，請不要猶豫與我聯繫。

此致，

Chapter 12

Recommendation Letter 2 （推薦函2）

Scenario（情境）：
推薦求職者之信件

To Whom It Concern: 2015-07-07

Firstly, I have to sincerely express that I have the honor and pleasure to send this recommendation letter for my partner, Dr. Allen Smith, whom I worked with in the past 5 years at iDigital Inc., Taichung, Taiwan.

As a supervisor of him, he undoubtedly demonstrated his unparalleled professional capabilities concerning the digital media design with insight and vision. His domain knowledge has tremendously contributed to our company with respect to the intangible asset.

Upon hearing his new career planning, I would like to strongly recommend Dr. Allen Smith although he has been extraordinarily precious during the continuous exponential growth of iDigital Inc.

If there is anything that you would like to confirm, please do not hesitate to contact with me during your tight schedules.

Cordially,

Becky Chen

Becky Chen

Professional Consultant

Research & Development Division

iDigital Inc., Taichung, Taiwan

Vocabulary or Phrases:

1. Undoubtedly 毫無疑問地
2. unparalleled 無與倫比的
3. tremendously 巨大地
4. contributed to 貢獻
5. with respect to 關於
6. extraordinarily 非凡地

主旨

敬要啓者：　　　　　　　　　　　　　　　　　　　　2015年7月7日

首先，我要真誠地表示，我很榮幸地為我的合作夥伴：Allen Smith博士——在過去5年曾與我於台灣的台中iDigital公司共事——發送此推薦信。

身為他的上司，他無疑地展示了他有關數位媒體設計無與倫比的專業能力、遠見與洞察力。他的領域知識大大地有助於我們公司的無形資產。

當聽到他新的職業生涯規劃，我想強烈推薦Allen Smith博士，儘管他對iDigital公司的持續倍數成長是非常有貢獻的。

如果有什麼事情你想確認，請不要猶豫在你緊湊的工作時間中與我聯繫。

真誠的，

Chapter 13 Application Letter（申請函）

Scenario（情境）：

Tim Wang想申請到Black Technical Enterprises上班

Tim Wang

No. 144, MinSheng Road, Taichung City, Taiwan, ROC

Tel: +886-4-2218-1122

Fax: +886-4-2218-1123

e-mail: Tim_Wang88@hotmail.com

Date: 2015-07-07

Becky Chen

Black Technical Enterprises

P.O. Box 123, Keeseville, NY, 12944, USA

Phone 518-555-2345

Dear Becky,

This is Tim Wang, who saw the job advertisement on your web site concerning the JOB-Code 345-1. I am extremely interested in this position. Consequently, I submit my resume accordingly.

Please kindly find the attached file, which is my CV. If my expertise in chemical engineering meets your requirement, please kindly contact me via the information in the CV.

I am willing to relocate if the one is part of the job requirements.

Best regards,

Tim Wang

Vocabulary or Phrases:

1. Consequently 因此
2. Accordingly 因此、根據

主旨

親愛的Becky：

我是Tim Wang，看到了你的網站上招聘代碼345-1的招聘廣告。我對這個職位非常感興趣。因此，我提出我的簡歷。

敬請參考附件的文件，這是我的簡歷。如果我在化學工程專業知識能滿足您的要求，請透過在簡歷中的信息聯繫我。

如果其中一個工作要求是搬遷住所，我也十分樂意。

誠摯的祝福，

Chapter 14 Job Interview Letter（工作面試函）

Scenario（情境）：
通知應徵者前來面試

Black Technical Enterprises

P.O. Box 123 • Keeseville • NY • 12944

Phone 518-555-2345

Dear Mr. Hank,

Up receiving your submitted resume, the recruiting committees were impressed by your exceptional credentials and solid background in academics.

Hence, we sincerely invite you for an interview with them. Please kindly notice that your interview time is 10:00 a.m., July 8, 2015 on the 6th floor of Black Technical Enterprises.

I am looking forward to hearing from you soon. We have reserved a parking space in B3 #89 for you, please notify the security guard and on drive directly through the gate on the day you arrived here.

Sincerely,

Becky Chen

Becky Chen

Director of human resources

Black Technical Enterprises

Vocabulary or Phrases：

1. impressed 給…極深的印象
2. exceptional 優秀的、卓越的
3. looking forward to 期待

主旨

親愛的Hank先生：

自從收到您所提交的簡歷後，招聘委員就對您傑出的憑證，以及堅實可靠的學術背景留下深刻的印象。

因此，我們誠摯地邀請您與他們進行面談。請注意，您的面試時間是2015年7月8日的上午10點，地點在Black Technical公司的6樓。

我非常期待您的回覆。我們為您保留了B3＃89的車位，您來的那天，請通知保全人員並直接通過大門。

誠摯的，

Chapter 15 Re-Scheduling Letter（重新預約函）

Scenario（情境）：
重新與對方敲定會面時間之信件

TO: Allen White

FROM: Becky Chen

DATE: 7 July 2015

SUBJECT: Meeting postponed

Dear Allen,

Suddenly, unexpected flaw of the semi-products occurred on the assembly line of SS-160. We are endeavoring to solve the problem to the best we could.

For the demonstration date of SS-160, which is two weeks from today, we are afraid that we can't meet your expectation. Hence, we have no choice but to postpone the meeting to 2:00 p.m., November, 3rd.

We sincerely apologize for any inconvenience that might occur due to the aforementioned issue.

Truly yours,

Becky Chen

Director of R&D

Black Technical Enterprises

Vocabulary or Phrases：

1. unexpected 不如預期的、不符合期望的
2. endeavoring 努力、盡力
3. demonstration 展示
4. have no choice but to 不得不…
5. inconvenience 不方便

主旨

親愛的Allen：

很突然地，半成品在SS-160的裝配生產線發生意外缺陷。我們正在盡我們全力努力解決這個問題。

關於SS-160的展示日，我們擔心我們不能達成您的期望。因此，我們別無選擇，只能推遲會議至11月3日下午2:00。

我們為上述問題可能發生的任何不便，真誠的道歉。

誠摯的，

Chapter 16 Sample Request Letter（樣品需求函）

Scenario（情境）：

請代工廠商提供樣品及交代運送方式與最後抵達日期

TO: Tiffany Kermert

FROM: Becky Chen

DATE: July 9, 2015

Hi, Ms. Kermert,

This is the confirmation of our sample request with regard to the following units:

1. SS-160 in its finished status with yield above 95%
2. SS-180 without the sealing process with yield above 98%

The production division needs them no more than August 10th. Please kindly airmail them and collect to:

Mr. Lowie Wei

iDigital Technical Enterprise

No. 144, MinSheng Road, Taichung City, Taiwan, ROC

Our company will disburse the funds within 30 days of shipment.

Best,

Becky Chen

Beck Chen

Vice President of R&D

Black Technical Enterprises

Vocabulary or Phrases:

1. with regard to 關於
2. yield 產量、收穫量、收益、利潤
3. no more than 不超過
4. disburse 支付、支出

主旨

嗨，Kermert小姐：

這是我們對於以下元件的樣品需求確認單：

1. SS-160完成狀態，良率95%以上
2. SS-180未密封狀態，良率98%以上

生產部門需要在8月10日前收到樣品。

請空運將它們寄給：

iDigital技術企業

中華民國台灣，台中市民生路144號

本公司將在發貨後30日內支付貨款。

敬啓，

Chapter 17 Relocation Letter（搬遷函）

Scenario（情境）：
公司搬遷昭告舊雨新知信函

iDigital Technical Enterprise

No.140, Minsheng Rd., West Dist., Taichung City 40306, Taiwan

Tel: +886-4-2218-1122

Fax: +886-4-2218-1123

9 July 2015

Re: Our new office location

Dear Customer,

It is our greatest pleasure that I am writing to inform you that on August 1st, we will open our new storefront in a more spacious and comfortable surroundings. The following is our new address without any changing in terms of contacting with us:

#288, Sec. 4, Taiwan Boulevard, Shalu Dist., Taichung City 43301, Taiwan.

The new location is right at the corner of the Pacific Department Store, where provides more parking facilities for our dearest customers.

Since we opened for business on August 1st, 2010, our business was running an exponential growth thanks to our customers' supports. During the grand opening, in order to show our deepest appreciation, 20% discount will be applied to every single product or service we provide.

As of August 1st, effective immediately, we will not be operating in the current location to provide the services. If you have any questions respecting the new location, please do not hesitate to call us and we'll be happy to assist you in any way we could.

We look forward to seeing you at our new location.

Sincerely,

Becky Chen

Becky Chen

Manager of Public relation

Enclosure if a map is enclosed

Vocabulary or Phrases：

1. grand opening 開幕儀式
2. appreciation 感謝
3. effective immediately 立即生效
4. respecting 關於

主旨

親愛的顧客：

這是我們最大的榮幸通知你們在8月1日，我們的新店面將在一個更寬敞和舒適的環境開張。以下關於我們的新地址以及我們的聯繫方式沒有改變：

43301台中市沙鹿區台灣大道4段288號

新的位置就在太平洋百貨的角落，在那裡我們提供更多的停車設施給親愛的顧客。

從我們開業於2010年8月1日時起，我們的業務營運指數的增長就得益於客戶的支持。為了表示我們最深切的感謝，在新店開幕時，將提供20%的折扣，可適用於每一個產品或服務。

截至8月1日後，我們將不會在目前所在地點提供服務。如果您有任何疑問關於新的地點，請不要猶豫打電話給我們，我們很樂意以任何我們能做到的方式來幫助您。

我們期待在新地點看到你們。

誠摯地敬上，

Chapter 18 Sick Leave Letter（病假函）

Scenario（情境）：

Tom Hank因病無法上班，需請病假

Dear Mrs. Becky,

I am writing this letter to notify you that I have to take one day sick leave due to the diarrhea last night. Therefore, tomorrow, July 8, I have to stay at home according to the diagnosis report from the physician.

I am awfully sorry that I have to be absent tomorrow, which I have three meetings to join. I will be more than happy to participate if video conferencing is available tomorrow.

Please kindly find the attached file, which is medical report from the clinic. Please feel free to let me now at your convenience should you have any inquiry concerning my sick leave request. I really appreciate your prompt attention to this matter.

Respectfully yours,

Tom Hank

Tom Hank

Vocabulary or Phrases：

1. Diarrhea 腹瀉
2. Therefore 因此
3. awfully 非常地
4. video conferencing 視訊會議

主旨

親愛的Becky女士：

我寫這封信來通知您，是由於昨晚的腹痛，我不得不請一天病假。因此，明天，7月8日，我將會根據來自醫生的診斷報告留在家裡。

非常抱歉，我不得不缺席明天的三個會議。如果明天能提供視訊會議，我會很樂意參加。

敬請參考附件文件，它是來自診所的醫療報告。請您隨意對我的病假請求做出任何詢問。我真的很感謝您對於此事的及時關注。

尊敬您的，

Chapter 19

Experienced French Instructor Wanted Letter（徵聘有經驗之法文教師廣告函）

Scenario（情境）：

公司要在報紙上刊登徵聘廣告

Experienced French Instructor Wanted

The Global Finance Company is recruiting a full-time French Instructor for its headquarter office in Taichung. The permanent position is suitable for anyone seeking a stable position in our company.

Requirements:

1. Bachelor's degree
2. Three or more years' teaching experience with adults is a must
3. Native speaker ONLY
4. Willing work over-time under pressure is a plus

Benefits:

1. Two weeks paid vacation on an annual basis
2. Monthly compensation is initially based on experience and education
3. Bonus from 1 to 5 months' salary
4. Housing allowance
5. Medical insurance
6. Work permit with ARC provided

All those interested candidates should e-mail their Curriculum Vitae (CV) to the contact person below. We sincerely apologize that only shortlisted ones will be notified.

Contact Person: Mrs. Wang

Tel: 04-2434-3456 Ext. 122 e-mail: wang0643@gmail.com

Global Finance Company

220, Sec. 3, Taiwan Boulevard, Taichung City 40301 Taiwan

Tel: +886-4-04-2434-3457 Ext. 113-115

service220@gmail.com

Vocabulary or Phrases:

1. Permanent 永久的
2. under pressure 壓力下
3. compensation 補償
4. Bonus 獎金
5. allowance 補貼

主旨

徵聘有經驗之法文教師

全球金融公司在台中的總部辦公室正在招聘一名全職的法國教師。長期的職位適合任何人尋求在我們公司有一個穩定的位置。

要求條件：

1. 大學學位
2. 須有三年以上與成人的教學經驗
3. 僅限以法語為母語的人
4. 願意在時間壓力下工作者更好

福利：

1. 每年兩週帶薪休假
2. 每月補貼最初是基於經驗和教育程度來給付
3. 1至5個月工資的額外獎金
4. 住房補貼
5. 醫療保險
6. 符合外國人居留證工作許可規定

所有對這些有興趣的人請寄電子郵件的簡歷（CV）給下面的聯絡人。我們真誠地道歉，只有入選的人會被通知。

Chapter 20 Signing MoU（簽署備忘錄）

Scenario（情境）：
兩所大學簽訂MoU（Memorandum of Understanding，備忘錄）之內容信函

A General Academic Cooperation Agreement
Between National Taichung University of Education (Taichung, Taiwan) and The Mongolian State University of Education (Ulaanbaatar, Mongolia)

Both universities agree to establish ties of friendship and cooperation for the purposes of promoting mutual understanding and academic (cultural) exchanges of faculties and students.

National Taichung University of Education (hereinafter referred to as NTCU) represented by President Szu-Wei Yang and the Mongolian State University of Education (hereinafter referred to as MSUE) represented by President B. Zhadambaa, mutually agree to promote scholarship activities and international understanding.

Objectives

The objectives of this agreement shall include but not be limited to the following:

1. The development of collaborative research projects involving our respective faculties.
2. The organization of joint academic and/or scientific activities, such as courses, conferences, seminars, symposia or lectures.
3. The exchange of research and teaching personnel.
4. The exchange of undergraduate and graduate students.
5. The exchange of academic publications and other materials of common interests.

Administration and Implementation

1. This general agreement will be in effect for a period of five (5) years at which time it shall be reviewed for possible extension.
2. Both institutions will designate an institutional coordinator to facilitate the development and conduction of the various activities.
3. The coordinators will be responsible for the direction and supervision of all activities of the joint programs under this agreement, subject to the rules and regulations pertaining to each university.
4. Either university may initiate specific proposals for activities through institution's coordinators, and upon signing by the President of MSUE and the President of NTCU, shall become part of the general agreement.
5. No financial obligations are assumed under this agreement. Funding for any specific program or activity shall be mutually discussed and agreed upon in writing by both parties prior to initiation of same.

6. NTCU offers one MSUE student per year to study at NTCU Chinese Language Center for paying 70% tuition fee (no longer than 1-year period).
7. The selection of students, faculties, and staff members participating in the exchange programs and projects will occur through a mutually agreed process between the two universities.
8. Supplemental agreements will address such issues as: specific duties and responsibilities, schedules, budgets, evaluations, and any other items necessary to efficiently accomplish the activity.

Approval of the General Academic Cooperative Agreement

This agreement will become effective upon being signed by the Presidents of both institutions. This agreement may be terminated by either party provided the terminating party gives written notice at least six (6) months prior to termination.

________________________	________________________
Dr. Szu-Wei Yang, President National Taichung University of Education	Dr. B. Zhadambaa, President The Mongolian State University of Education
Date:________________	Date:________________

Vocabulary or Phrases:

1. Hereinafter 在下
2. Implementation 履行、完成
3. Designate 標出、表明、指定
4. Coordinator 同等的人（或物）、協調員
5. subject to 受制於、以……爲條件
6. financial obligations 財務責任
7. mutually 互相地

主旨

國立台中教育大學（台中）和教育蒙古國立大學（蒙古烏蘭巴托）的全面學術合作協議

兩所大學同意建立友好合作關係，以促進院系和學生間的相互理解和學術（文化）交流為目的。

國立台中教育大學（以下簡稱NTCU）校長泗偉陽教育蒙古國立大學（以下簡稱MSUE）校長B. Zhadambaa為代表，雙方同意促進學術活動和國際交流。

目標

這個協議應包含但不限於以下內容：

1. 關於我們各自院系所合作的研究項目的開發。
2. 聯合學術和／或科學活動，如課程、會議、研討會、專題討論會或講座的組織。

3. 研究和教學人員的交流。

4. 本科生和研究生的交流。

5. 學術刊物和共同關心的其他資料的交換。

管理與實現

1. 本協議會生效為期五(5)年，在此期間經審查後可能延長。
2. 這兩個機構將指定一個機構協調，以促進各項活動的發展和傳達。
3. 協調員將根據本協議負責指導和監督聯合方案的所有活動，遵守屬於每所大學各項規章制度。
4. 只要是大學可透過機構的協調進行的各項活動的具體建議，並由MSUE校長和NTCU校長所簽署的，都將成為總協議的一部分。
5. 根據本協議沒有財政承擔義務。資助任何具體計畫或活動之前，應由雙方共同討論並取得書面同意。
6. NTCU提供每年一次MSUE學生在NTCU中國語言中心學習，並支付70%的學費（不超過1年）。
7. 獲得兩所大學之間相互同意的學生、教職員和工作人員，可選擇參與交流計畫和項目。
8. 補充協議將解決以下問題：特定的義務和責任、進度、預算、評估，並有效地完成該活動的任何其他項目。

核准一般性學術合作協議

該協議將經雙方校長簽字而生效。該協議的終止，須由任何一方至少於六(6)個月前提出終止合約的書面通知才能終止。

Module

03

Special Occasions 特殊場合

Module 03

Chapter 1 Anniversary Letter（週年紀念函）

Scenario（情境）：
祝賀別家公司15週年紀念

Dear Sophia,

We are so excited to be notified by your invitation letter to join your 15 years anniversary since you launched the business. Demonstrably, your company had established the role model with unparalleled achievements in this business arena in the past 15 years. President Lowie will show up in person on behalf of our company.

As a close partner of you in the supply chain of contemporary semiconductor industry, it goes without saying that you are one of the most promising corporates in the private sectors.

Your continuous profitable business model embodies the insight and vision of your company, 15 years in a row, our excitements for your company is beyond description.

Truly,

Becky Chen

Senior Vice President

iDigital Technical Enterprise

No. 144, MinSheng Road, Taichung City, Taiwan, ROC

e-mail: hr@idigital.com.tw

Tel: +886-4-2218-1122

Fax: +886-4-2218-1123

Vocabulary or Phrases:

1. Demonstrably 明確地
2. Unparalleled 無與倫比的
3. on behalf of 代表
4. contemporary 現代
5. insight 洞察力
6. vision 願景、遠景
7. in a row 連續
8. beyond description 無法形容的

主旨

親愛的Sophia：

我們都非常興奮被通知能參與您開業15週年的紀念會。明確地說，您的公司在過去15年來，以無與倫比的成就於這商業界樹立了榜樣。Lowie主席將代表本公司親自出席。

身為您當代半導體工業供應鏈上緊密的夥伴，您是私營企業中最具前景的合作伙伴之一，這件事可說是不言而喻。

您持續獲利的商業模式體現了貴公司的洞察力和展望，連續15年，我們對貴公司的興奮之情溢於言表。

真誠的，

Chapter 2 A New Born Baby Letter（嬰兒誕生函）

Scenario（情境）：
慶祝同仁或朋友剛獲得一位新生寶寶

Black Technical Enterprises

211 Main Street • Keeseville • NY • 12944

Phone 518-555-2345

July 10, 15

Congratulations for having a newborn baby

John Chen

513, Sec. 1, Taiwan Boulevard,

Taichung City, 43301 Taiwan

Dear John,

Upon hearing that you and Nancy have the newborn baby last night, I am excited and absolutely have to send this letter immediately to share your happiness.

All persons in our office are also happy by this breaking news. We all believe that you and Nancy are going to be much more busier than usual due to this precious gift from the God. Definitely, your family is going to have plenty of joy due to the new member.

Please kindly post the photos on the Facebook whenever you and Nancy have spare time. We are looking forwards to seeing those wonderful pictures on line.

Lastly, we deeply wish Nancy and the newborn baby are doing very well.

Truly yours,

Becky Chan

Becky Chan

Manager of Human Resources

Black Technical Enterprises

Vocabulary or Phrases:

1. absolutely 絕對、完全
2. Definitely 無疑地
3. Lastly 最後

主旨

親愛的John：

自從昨晚耳聞你與Nancy喜獲新生兒之後，我就非常興奮且一定要馬上寄此信來分享你們的喜悅。

我們辦公室裡的所有人，也都為這個驚人的消息而開心不已。我們都相信你跟Nancy將為這個上天賜給你們的珍貴禮物，而比平常更加地忙碌。無疑地，你們家將因這個新成員而擁有更多的歡樂。

當你與Nancy有空閒的時候，請把照片發布在臉書上。我們都很期待在網路上看見那些美好的照片。

最後，我們深深地希望Nancy和新生兒一切安好。

真誠的，

Chapter 3 Apology Letter（致歉函）

Scenario（情境）：因生產線出問題無法如期交貨

TO: Joyce Smith

FROM: Becky Chen

DATE: 10-Jul-15

SUBJECT: Apologize for the delay of your shipment

Dear Ms. Smith:

Firstly, I have to sincerely apologize that your purchase order # 144-65 will be two weeks delayed due to the unexpected malfunction of our assembly line because our consulting company was lunching the business process reengineering, which supposedly will increase the yield and shorten the time to market. Unfortunately, it needs tune up a little bit.

We fully understand that this will cause you extreme inconvenience due to the supply chain management in your industry. We will pay you the penalty based on our contract before the due date.

In an effort to show our apology, we will offer 20% off next time when you put the purchase order. We look forwards to continuing servicing your business and hope you can tolerate the delay.

Vocabulary or Phrases:

1. Firstly 首先
2. unexpected 未預期的
3. business process reengineering 業務流程重組
4. inconvenience 不便
5. supply chain management 供應鏈管理
6. penalty 罰款

主旨

親愛的Smith小姐：

首先，因為我們諮詢公司進行業務流程重組，以期增加產量並縮短送達市場的時間，造成未預期的組裝線失靈，導致你購買的型號# 144-65將延遲兩星期抵達，我必須誠摯地為此致歉。不幸地，它還需要些微地調整。

我們完全了解這將造成您產業中，供應鏈管理極大的不便。我們將在截止日前，依據我們的協約賠償您。

為了向您表達我們的抱歉之意，下次您下訂單時，我們將提供八折的優待。我們期望能繼續為您的事業服務，並希望您能容忍此次的延遲。

Condolence Letter（弔唁函）

Scenario（情境）：
公司向往生的同仁家屬致哀

July 11, 15

Mrs. Jenny Wang

288, Northern Boulevard

Old Westbury, NY 11568-6000, USA

Dear Jenny,

I have no words to express how I am deeply sorry hear about the news respecting the sudden death of your husband, and I want to express my deepest sympathy to you and your family on behalf of iDigital Technical Enterprise.

Your husband was an industrious and professional project leader in our company for many years. Unquestionably, his pass away is the greatest loss of our company's intangible assets. In addition, he left many colleagues here, who were all his close friends.

There are no words to express the heartache we feel for your family. Our love thoughts and prayers are with you and your entire family. As always, we are deeply impressed by your husband's amazing spirit, long lasting strength, and decent personality.

Sincerely yours,

Becky Wang

Becky Wang

President

iDigital Technical Enterprise

No. 144, MinSheng Road, Taichung City, Taiwan, ROC

e-mail: hr@idigital.com.tw

Tel: +886-4-2218-1122

Fax: +886-4-2218-1123

Vocabulary or Phrases:

1. deepest sympathy 最深切的同情
2. on behalf of 代表
3. industrious 勤勞
4. Unquestionably 毫無疑問
5. pass away 去世
6. intangible assets 無形資產
7. impressed 印象深刻

主旨

親愛的Jenny：

聽聞您丈夫突然過世的消息，我真的沒有適合的文字來表達我深切的遺憾之意，同時我也代表iDigital科技公司對您以及您的家人表達最深的同情。

您的丈夫是我們公司多年以來，一位勤勞且專業的專案領導人。毫無疑問地，他的離開是我們公司無形資產最重大的損失。此外，他在這裡留下的許多同事，他們都是他最親近的朋友。

沒有任何言語可以表達我們心中對您的家庭的悲慟。我們充滿愛的意念和祈禱，將伴著您與您的整個家庭。如同往昔一般，我們對您的丈夫令人驚異的靈魂、經得起考驗的實力以及良好的人格永遠懷念。

您忠誠的，

Chapter 5 Farewell Letter（告別函）

Scenario（情境）：
本人將離職，向相關同仁發告別信函

TO: All employees of Black Technical Enterprises

FROM: Tommy Lin

DATE: 2015-07-11

SUBJECT: Farewell Letter

Dear colleagues,

This is Tommy Lin, who joined the company 4 years ago with unforgettable working experiences with all of you. I have to announce that starting on 2015-12-1 I will be no longer servicing the company due to my personal career planning.

I cherish the moment that I had with all of you in every single project we fulfilled. Unspeakably, I have the honor working for this company as well as working with all of you.

Additionally, I will be relocating to City of Los Angels, where my fiancée stays. Since my company e-mail account will be terminated soon, tommy23@hotmail.com will the alternative to reach me.

Welcome to City of Los Angels, where I will definitely host all of you when you visit me.

Sincerely yours,

Tommy Lin

Tommy Lin

Manager of production division

Black Technical Enterprises

Vocabulary or Phrases:

1. Colleagues 同事
2. Unforgettable 難忘的
3. Unspeakably 無法形容
4. Additionally 此外
5. Fiancée 未婚妻
6. Terminated 終止
7. Definitely 絕對

主旨

親愛的同事們：

我是已經在這個公司待了四年，並與你們一同留下許多難以忘懷的回憶的Tommy Lin。我在此聲明從2015年12月11日開始，基於個人職業生涯規劃，將不再於此公司服務。我珍惜與你們每個人一起完成每個企劃的分秒。無法用言語形容，我竟擁有這個榮幸為這個公司工作並與你們共事。

此外，我將移居到我未婚妻所居住的城市洛杉磯。由於我的公司郵件帳號將很快就會被終止，tommy23@hotmail.com這個郵件帳號將可替代前者連繫到我。

歡迎來洛杉磯，當你們來拜訪我時，我絕對會作東款待你們的。

你們忠誠的，

Chapter 6 Negotiation Letter（談判函）

Scenario（情境）：
公司開出薪資條件不如預期，希望資方有轉寰空間

Mr. John Morris

Black Technical Enterprises

P.O. Box 123 • Keeseville • NY • 12944

Phone 518-555-2345

hr@balcktech.com

Dear Mr. Morris,

I really appreciate your job offer in regard to the developer of 3D animation in your esteemed corporation. On the other hand, I am extremely interested in this position as well as joining your outstanding R&D division.

Frankly speaking, I have other job offers simultaneously in the similar fields. I would like to humbly discuss with you concerning the possible raise in my annual compensation from 78k to 95k. Please kindly refer to my CV, which pinpoints my 10 years experiences in this arena. Additionally, the projects I have done are worthy of demonstrating in contemporary multimedia communication channels.

Definitely, every company has its policies and undoubtedly I will be waiting for your kind feedback. If we need more mutual understandings, I would be more than happy to make an appointment with you for detailed discussion if you feel feasible.

I will be looking forwards to receiving your quick response.

Best regards,

Becky Chen

Becky Chen

Vocabulary or Phrases:

1. in regard to 關於
2. esteemed 受人尊敬的
3. frankly speaking 老實說
4. simultaneously 同時地
5. humbly 謙卑地、虛心地
6. compensation 補償
7. pinpoints 明顯指出
8. are worthy of 值得
9. definitely 絕對地
10. undoubtedly 毫無疑問地
11. feasible 可行的、可允許的

主旨

親愛的Morris先生：

我很感謝您提供我在貴公司擔任3D動畫的開發員。另一方面，我對這個職位以及加入您優秀的研發部門都非常感興趣。

老實說，我在同時也收到了其他類似領域的工作邀請。我希望能夠虛心地與您討論將我的年薪從78 K 提升至95 K 的可能性。請參考我的簡歷，其中指出了我在這個領域擁有10年的資歷。除此之外，我所執行過的計畫在當代的多媒體通信管道都是值得成為典範的。

當然，每間公司都有自己的政策，毫無疑問地，我會耐心地靜候您的回覆。如果您認為可行，認為我們需要取得更多的共識，我也很樂意安排與您碰面商討細節。

致上最好的祝福，

Chapter 7

Conflict Resolution Letter （衝突解決信函）

Scenario（情境）：
當訂單與對方有認知上落差時之溝通信函

TO: Joyce Smith

FROM: Becky Chen

DATE: 10-Jul-15

SUBJECT: Conflict Resolution of PO #334-56

Dear Ms. Smith,

We delivered your PO #334-56 last week and I just got the confirmation from you that the package has been firmly received. Unfortunately, we got the fax this morning from you indicating that the defect rate is above your expectation.

We got the confirmation from the QC department verifying that the yield of that package was above 95%. It looks like that we have a conflict of recognition regarding the quality of that package.

We strongly recommend that you seek the third party to verify the quality of the package and we will absolutely pay for the inspection fee. If the testing result from the third party showing the defect rate is still below 95%, we will reimburse the deposit to you immediately.

I sincerely apologize if there is any misunderstanding of this PO. We will be waiting for your further information even if you decide to return the package without the inspection from the third party. We are willing to provide the services upon your requests.

Sincerely yours,

Becky Chen

Becky Chen

Vocabulary or Phrases:

1. unfortunately 非常不幸地
2. defect rate 缺陷率、不良率
3. yield 良率
4. conflict of recognition 認知衝突
5. the third party 第三方
6. absolutely 絕對地
7. reimburse 補償
8. deposit 保證金

主旨

親愛的Smith小姐：

我們上週寄送了您的定單#334-56，而我剛才收到了來自您的確認信，說明包裹已經確實送到。非常不幸地，我們今早收到了您的傳真，表示商品的不良率超過了您的預期。

我們取得了來自品質檢驗部門的確認書，證實該批產品的良率仍然是高於95%的。這樣看來，我們似乎對於該批產品的品質有認知上的衝突。

我們強烈地建議您尋找第三方來鑑定這批貨的品質，而我們理當會支付檢驗的相關費用。如果第三方的檢驗結果顯示良率仍然低於95%，我們會立即將保證金退還給您。

如果您對於本次訂單有任何的誤解，本人在此致上誠摯的抱歉。我們將會等待您接下來的消息，即使您決定直接將商品退回而不經過第三方的檢驗。我們仍願意根據您的要求提供相對的服務。

您忠誠的，

Unexpected Accidents / Events Letter（意外事件 / 事故信函）

Scenario（情境）：
員工發生意外，撰寫慰問信

John Chen

Apartment 3B, #23 Main Street

Rocky, NY, 12456

Dear John,

Upon hearing your unexpected car accident, we deeply feel sorry about that issue. Please do not worry about the road show next week, we will have alternative in case you do not feel comfortable to be present. Furthermore, Steve will be temporarily substituted your position until you come back to the company when you are fully recovered.

Since this is the 2nd car accident in this company this month, The CEO, Helen, decides to provide the shuttle bus running between the station and the company only for our company employee since November 1st, effective immediately. The bus schedules will be announced momentarily.

Please do not worry about the errands in the office and we hope to you back in good shape whenever you feel comfortable. You have our best wishes and please take care.

Truly,

Kevin Bruce

Kevin Bruce

Manager, Human Resources

Black Technical Enterprises

P.O. Box 123 • Keeseville • NY • 12944

Phone: 518-555-2345

Email: hr@blacktk.com

Vocabulary or Phrases:

1. unexpected 出乎意料的
2. road show 巡迴參展
3. alternative 替代
4. effective immediately 立即生效
5. errands 雜務

主旨

親愛的John：

在聽見你車禍之後為你感到非常地遺憾。請別擔心下週的巡迴參展，如果你不便參與，我們會有個替代方案。主要是Steve會暫時替代你的位置，直到你完全康復回到公司。

由於這是公司這個月發生的第二件車禍，執行長Helen決定從11月1日起，在公司與車站之間提供專門為公司員工服務的接駁車，且立即生效。接駁車時間表將在近期內公布。

請別擔心辦公室內的雜務，我們希望你在健康的情況下回歸，祝早日康復。

真誠的，

Module

04

Quotations, Outsourcing, and other e-Commerce Occasions
報價、委外、其它電子商務場合

Module 04

Chapter 1 RFQ (Request for Quotation)（詢問報價信函）

Scenario（情境）：
要求美國公司限期報價

25th September, 2016

Mrs. Helen Swan

Black Technical Enterprises

211 Main Street • Keeseville • NY • 12944

Phone 518-555-2345

OBJECT: INVITATION TO QUOTE PRICE GOODS

Dear Helen,

We are very interested in purchasing your SS-160 from on-line catalogue.

Please kindly quote your ordinary unit price for supplying these goods as well as your discount for voluminous procurement.

Please also indicate the followings:

1. We would like to purchase an annual quantity of 20,000 to 30,000 units.

2. FOB Taiwan

3. Please quote the prices including the sale taxes.

4. Please indicate the accurate delivery time once you receive our PO.

5. The costs for delivery must be included in your quotes.

6. Terms of payments.

All price quotations must be precise along with their expiration date. We expect to receive your price quotations no later than 31st October, 2016.

If you have any confusion, please do not hesitate to contact with me via the following information.

Sincerely,

Becky Wang

Becky Wang
Manager, Division of Purchasing
iDigital Technical Enterprise
No. 144, MinSheng Road, Taichung City, Taiwan, ROC
e-mail: pr@idigital.com.tw
Tel: +886-4-2218-1122
Fax: +886-4-2218-1123

Vocabulary or Phrases：

1. voluminous 大量的
2. procurement 採購
3. FOB (Free on Board) 船上交貨
4. expiration 截止

主旨

親愛的Helen：

我們對於購買您線上目錄中的SS-160很感興趣。煩請您告知商品經過大量採購的折扣後的單價。

請看下列說明

1. 我們希望每年能夠購買20,000到30,000的數量。
2. 船上交貨。
3. 請在單價的標示上包含銷售稅。
4. 一旦您接下訂單時，請告知準確的到貨時間。
5. 運輸費用必須包含在您的開價中。
6. 付款條件。

所有的報價單必須準確地附上截止日期，我們期望在2016年10月31日之前收到您的報價單。

如果有任何疑問，請立即以下列資訊聯繫我們。

誠摯的，

Chapter 2 Place an Order（下訂單函）

Scenario（情境）：
向賣方下訂單之書函

Black Technical Enterprises

211 Main Street • Keeseville • NY • 12944

Phone 518-555-2345

call_center@blacktech.com

PURCHASE ORDER

NUMBER: 001/02 DATE: Dec. 15th, 2016

Black Technical Enterprises has this day bought the undermentioned goods from Global Cable Ltd. to be delivered in good order subject to the terms and conditions stated hereunder, unless otherwise specified:

DESCRIPTION	QUANTITY	UNIT PRICE	AMOUNT
P/N 2345-1 Cable, as last delivered	1,000 PCS	USD$1.891	USD$1,891.00

SHIPPING DOCUMENTS: Please list Black Technical Enterprises as the buyer and consignee.

DESTINATION: Black Technical Enterprises
c/o Cruiser Brokers
218 Main Street

Keeseville • NY • 12944

Attn: Bruce Lin

DELIVERY: ASAP

PACKING: For air freight-ship freight collect

QUALITY: Same as last purchase

REMARKS: JACK Technical Services to be the agent of Black Technical Enterprises for all technical questions, quality inspection, communication, and shipping information

Please acknowledge receipt and acceptance of this order as soon as is convenient by returning the copy duty signed.

Confirmed and Accepted by	Buyer's Signature
	John Lander
Ellen Smith	John Lander
(Global Cable)	(Black Technical Enterprises)

Vocabulary or Phrases:

1. PCS (pieces的縮寫，如果單數，則使用PC)
2. c/o (care of/ in care of) 由……轉交
3. freight collect 運費由買方貨到付款

主旨

訂單

號碼: 001/02　　　　　　　　　　　　　　日期: 2016年12月15日

Black Technical Enterprises從Global Cable有限公司，買了下列的商品，公司交付的訂單受到本協議規定，除非另有規定的條款和條件：

敘述	數量	單價	金額
P／N 2345-1 Cable, 如上次訂單	1,000 件	美金$1.891	美金$1,891.00

運送資料：請將Black Technical Enterprises列為買方和收貨人。

目的地：Black Technical Enterprises 研究公司

轉交由Cruiser Brokers

218 Main Street

Keeseville • NY • 12944

Attn: Bruce Lin

交貨：越早越好。

包裝：空運－船運費到付。

品質：如上次訂單。

備註：JACK Technical Services為Black Technical Enterprises的顧問，代為回答所有技術上問題、品質檢定、溝通和運送詳情。

請在收到訂單並確認後，盡速送回責任簽署。

Chapter 3 Material Requirement Planning（物料需求規劃信函）

Scenario（情境）：
因零組件供應鍵需求，請供應商補貨

Mrs. Jenny Lowie

Black Technical Enterprises

211 Main Street • Keescville • NY • 12944

Phone 518-555-2345

Dear Jenny,

Many thanks for your letter of June 21st and apologies for taking so long to reply due to the enormous backlog of e-mails.

To be frank, we are going to run out of SS-160, which you are our major component supplier, in two weeks. Our purchase order can be seen to the end of this year. Therefore, we would like to seriously consider signing a 6 months contract with you concerning the un-interrupted, steady, and high quality supply regarding SS-160 components.

It goes without saying that December is the busiest season of the year and the aforementioned scenario of supply is our rule of thumb in order to meet our Material Requirement Planning. If it is hard for you to fulfill the obligation of our cooperation, please kindly let me know as soon as possible. We deeply appreciate your quick response.

Best wishes,

Becky Wang

Becky Wang

Manager, Division of Purchasing

iDigital Technical Enterprise

No. 144, MinSheng Road, Taichung City, Taiwan, ROC

e-mail: pr@idigital.com.tw

Tel: +886-4-2218-1122

Fax: +886-4-2218-1123

Vocabulary or Phrases:

1. Backlog 存貨、累積
2. To be frank 坦白說
3. It goes without saying 不用說
4. rule of thumb 經驗法則、基本原則

主旨

親愛的Jenny：

非常感謝您6月21日的信，以及為處理大量電子郵件的累積而延遲回復的道歉。

坦白說，我們的SS-160快用完了，而您是我們近兩週的主要零件供應商。今年年底我們的採購訂單就會到期，因此，我們想認真地考慮與您簽下六個月的合約，以達成不中斷、穩定且優質的SS-160組件供應鏈。

不用說，12月份是一年中最繁忙的季節，上述的供應情況是我們為了滿足我們的物料需求所計畫的基本原則。如果這對您來說是很難達成的合作義務，請您盡快讓我知道。我們深深感謝您的快速回復。

誠摯的祝福，

Chapter 4

Manufacturing Resource Planning（製造資源規劃信函）

Scenario（情境）：
詢問是否有符合新需求的機器

TO: Allen White

FROM: Becky Chen

DATE: 24-Jul-15

SUBJECT: urgent MRP request

Dear Mr. White,

It has been a while that we did not contact with each other. We deeply wish that everything goes well of your corporation.

Concerning the packaging machines we bought from you last year, we would like to add other functions to make the production of our assembly line more effective and efficiency.

We need the packing machine to read the QR code in addition to the existing capability of scanning bar code. If you can modify the machines or add some modules to fix the problems we encounter, please kindly send us the quotation of enabling that function. We expect to see the quotation within 3 weeks if it does not bother you too much.

If you have further questions, please directly contact with my assistant, Mr. Bruce Lam (bruce_lam34@gmail.com), who will be the contact window respecting the above issue.

We are looking forwards to hearing from you soon.

Sincerely yours,

Becky Wang

Becky Wang

Manager, Division of Purchasing

iDigital Technical Enterprise

No. 144, MinSheng Road, Taichung City, Taiwan, ROC

e-mail: pr@idigital.com.tw

Tel: +886-4-2218-1122

Fax: +886-4-2218-1123

Vocabulary or Phrases:

1. It has been a while 已經有一段時間了
2. effective 有效的
3. efficiency 效率

4. in addition to 除了…之外

5. modules 模組、組件

主旨

親愛的White先生：

已經有一段時間，我們沒有與對方聯繫了。我們深切希望貴公司順利且安好。

關於我們去年向您買的包裝機，我們想增加其他功能，好讓我們的生產線更有效益和效率。

我們所需要的包裝機除了現有解讀條碼的功能外、還要能解讀QR code。如果您可以修改的機器或者添加一些模組來解決我們遇到的問題，請您向我們提出實現該功能的估價。我們預計3週內報價，希望不會打擾您太多。

如果您還有其他問題，請及時與我的助手，布魯斯林先生（bruce_lam34@gmail.com）聯繫。

我們靜待您的佳音。

誠摯的祝福，

Chapter 5 Adjust Quotation Letter（調整報價函）

Scenario（情境）：
向對方進行報價

iDigital Technical Enterprise

513, Sec. 1, Taiwan Boulevard,

Taichung City, 43301 Taiwan

July 10, 15

Peter Ian

Black Technical Enterprises

211 Main Street • Keeseville • NY • 12944

Phone 518-555-2345

Dear Peter,

Thank you for your reply concerning the acceptance of quotations and terms on July 8, 15. I have to sincerely apologize that we made a mistake on that quotation. Unfortunately, the new quotation is as followings:

Set	Model	Price per Set	Warranty
SS-160	Malaysia	$ 2.33 USD	1 year
SS-161	Czech	$ 1.56 USD	3 month
SS-162	Japan	$ 2.88 USD	1.5 years

Our normal trade discount is 10% for 15 days and 7% extra if PO is made for more than 3,000 sets at a time.

Since stock level is pretty tight in recent 3 months, we will provide the products to our customers based on first-come-first-serve policy.

Once again, please kindly accept our typo in last quotation, which is seldom occurred in our operations.

Again, thank you for your understanding and we hope to see the PO from you soon.

Kind regards,

Becky Wang

Becky Wang

Manager, division of warehousing

iDigital Technical Enterprise

513, Sec. 1, Taiwan Boulevard,

Taichung City, 43301 Taiwan

e-mail: hr@idigital.com.tw

Tel: +886-4-2218-1122

Fax: +886-4-2218-1123

Vocabulary or Phrases：

1. Unfortunately 不幸的
2. first-come-first-serve 先來先得
3. typo 打字錯誤

主旨

親愛的Peter：

謝謝您2015年7月8日的接受回復與報價。關於報價的錯誤，我要誠摯地向您道歉。令人遺憾的是，新的報價如下：

型號	原型	單筆報價	保證期
SS-160	馬來西亞	$ 2.33 USD	1 year
SS-161	捷克	$ 1.56 USD	3 month
SS-162	日本	$ 2.88 USD	1.5 years

我們的正常貿易優惠是15天10%以及額外的7%優惠，如果訂單單筆超過3000 的話。

由於近3個月庫存水準是相當吃緊的，我們將根據先來先服務政策，提供的產品給我們的客戶。

再次，請您原諒我們報價的打字錯誤，這在我們的業務中是很少發生的。

再次感謝您的理解，我們希望能盡快看到您的訂單。

誠摯的問候，

Chapter 6 Follow-Up Letter（追蹤函）

Scenario（情境）：

面試完後，表達謝意與表示自己十分有意願加入公司

John Chen

513, Sec. 1, Taiwan Boulevard,

Taichung City, 43301 Taiwan

Tel: +886-4-2245-4578

July 25, 2015

Manager Becky Wang

iDigital Technical Enterprise

No. 144, MinSheng Road, Taichung City, Taiwan, ROC

e-mail: hr@idigital.com.tw

Tel: +886-4-2218-1122

Fax: +886-4-2218-1123

Dear manager Wang,

It was a great pleasure meeting you on last Tuesday. I really appreciate the time you spent from your tight schedules to interview me. In addition, I felt the whole research team is amazing and it is entirely beyond my imagination.

As I stated in my CV, my specialty is 3D design, which is consistent with the vision of your esteemed company. This is the organization that I have dreamed for my career planning. If there is a match, I would like to devote myself to help the corporate grow to the best I could.

Once again, I would like to express my deepest appreciations to all the staffs who interviewed or assisted me on that day. If you have further questions, please do not hesitate to contact with me.

I am looking forwards to hearing from you in the near future.

Sincerely,

John Chen

Vocabulary or Phrases:

1. beyond my imagination 超出我的想像
2. specialty 專長
3. consistent 持續的、一致的
4. esteemed 令人尊敬的
5. devote myself to 爲…貢獻一己之力

主旨

親愛的王經理：

在上週二見到您是我莫大的榮幸。我真的很感謝您從緊湊的行程中花時間來為我面試。另外，我覺得整個研究團隊是驚人的，它完全超出了我的想像。

正如我在簡歷中所說，我的專業是3D設計，這與貴公司的願景是一致的。這是我職業生涯規劃中夢想的組織。如果有機會，我想貢獻一己之力於幫助貴企業成長。

再次，我想對所有在那天面試或協助我的工作人員表達深深的謝意。如果您還有其他問題，敬請與我聯繫。

我期待在不久的將來能聽到您的好消息。

誠摯的祝福，

Chapter 7 A thank you Letter 1（感謝函1）

Scenario（情境）：
給合作公司主管的感謝函

Black Technical Enterprises

P.O. Box 123 • Keeseville • NY • 12944

Phone 518-555-2345

January 8, 2016

Kevin Lai

651, Sec. 2, Taiwan Boulevard, Taipei City, 45302 Taiwan

Dear Mr. Kevin,

Thank you again for the dinner and the opportunity you provided to me that the combination of production chain between our companies. I've attached a file highlighting the details of work assignments in our company.

As I mentioned to you during our meeting, I believe the cooperation between two companies is beneficial and would bring promising future in the near future.

I'm looking forward to the quick response from you if possible. Please don't hesitate to consult with me if you have any problem arises.

Thank you again.

Best wishes,

Candy Wu

Director of diplomatic section

Black Technical Enterprises

Vocabulary or Phrases:

1. combination 組合、合作
2. production chain 生產鏈
3. hesitate 猶豫
4. consult 請教、諮詢

主旨

親愛的Kevin先生：

再次感謝您的晚餐與您所提供的與貴公司生產鏈合作的機會。以下附上說明我們公司內部工作分配的檔案。

正如我於我們會議中所提及的，我相信我們兩家公司的合作是有利可圖的，且在不久的將來能帶給我們有前景的未來。

如果可以的話，我期待您快速的回復。若有任何疑問，還請不要猶豫地與我洽詢。

再次感謝您。

獻上最好的祝福，

Appreciation Letter 2（感謝函2）

Scenario（情境）：
對公司獲得採用表示感謝

Dear Mr. Smith,

Thank you for choosing J.ZON Advertising to assist a presentation on the new product next month. We had been specialized in this area for more than 20 years. Base on the experience and professional competences we possess, you can count on us.

The attached file is the preliminary conception of new product advertising project we had imagined. If you have any suggestion or question, please do not hesitate to let us know. In addition, it would be great if you could spare some time to let us show you some details.

Again, thank you for choosing us. We are looking forward to establishing a long and prosperous relationship with you in the future.

Sincerely,

Stephen Lu

Stephen Lu

Manager, Marketing Department

J.ZON Advertising Enterprise

Phone 518-555-2345

Vocabulary or Phrases:

1. competence 能力
2. count on 信賴、指望…
3. preliminary conception 初步構想
4. In addition 此外
5. prosperous 繁榮的

主旨

親愛的Smith先生：

感謝您選擇J.ZON廣告公司協助下個月的新產品發表會。我們從事這個專業領域已有超過20年的歷史。以我們所擁有的豐富經驗以及專業能力，您完全可以放心將這個重責大任交予我們。

附件的內容是我們對於新產品廣告企劃的初步構想，如果您有任何建議或疑問，請不要猶豫讓我們了解。此外，如果您能夠撥出一些時間讓我們向您說明一些企劃的細項，我們將會感到萬分榮幸。

再次感謝您選擇我們公司，我們期待與您在未來建立一個長期而良好的關係。

誠摯的，

Chapter 9

Congratulation Letter（祝賀函）

Scenario（情境）：
祝賀新公司成立並尋求合作

To: Patty Mills

From: Danny Green

Date: 5/20

Subject: Congratulate We on the Opening of a New Business

Dear Mr. Mills,

Congratulations on opening your second business. It has just come to our attention that you have lately opened your new company in Texas. This distinguished accomplishment says a lot regarding customer satisfaction and the quality of your products and services several years in a row.

As you know, our companies have had a long business association with some organizations in the USA. We look forward to collaborating with you with respect to your newly established company. Joint venture is one possibility for two of us to consider in the pioneering research fields. Please let us know if we can be of any assistance to you. We will be delighted to provide any assistance. We wish you the very best of luck and a prosperous future.

Sincerely,

Danny

Danny

Vocabulary or Phrases：

1. distinguished 傑出的
2. accomplishment 成就
3. regarding 關於
4. collaborate 合作
5. Joint venture 合資企業
6. pioneering 先驅的
7. assistance 幫助
8. prosperous 繁榮的

主旨

親愛的Mills先生：

恭喜您開了第二家公司，我們注意到您新開在德州的公司。這傑出的成就也說明了這些年來客人對你們公司的滿意度，還有良好的商品品質及服務態度。

正如您所知的，我們一直與美國的公司有商務往來。我們非常期待能與您新成立的公司合作。且我們相信，透過雙方的合資，將能夠使我們獲取此一領域的先驅者地位。請讓我們知道有什麼是我們可以幫得上您的。我們非常樂意提供您協助。祝您好運且前程似錦。

誠摯的，

Module

05

Business Traveling / Road Show
商務旅行／商務參展

Module 05

Airplane Ticket Reservation（飛機票預約函）

Scenario（情境）：
向旅行社預訂機票

Sherry Lin

EZ Travelling Inc.

No. 233, Peace Avenue, Taipei City, Taiwan, ROC

email: sherry_lin@eztravelling.com.tw

Tel: +886-4-2218-1122

Fax: +886-4-2218-1123

July 13, 2015

Dear Mrs. Lin,

This is Becky from iDigital Technical Enterprise. Please kindly reserve two round-trip tickets to Hong Kong on this Saturday morning as early as possible. President Lee and I have to take a business trip for the road show. You have our passport data as usual.

Please keep the returning ticket open since we do not know the exact time to come back. Both seats have to be aisle seats. No upgrade is necessary for both of us.

Please just e-mail the e-ticket to my e-mail address: becky068@hotmail.com.

Cancellation policies might be applied according to the contract we signed due to the fact that we are running a very tight schedule at this quarter.

Best,

Becky Chen

Senior Vice President

iDigital Technical Enterprise

No. 144, MinSheng Road, Taichung City, Taiwan, ROC

e-mail: hr@idigital.com.tw

Tel: +886-4-2218-1122

Fax: +886-4-2218-1123

Vocabulary or Phrases:

1. round-trip 往返
2. returning ticket open 不定期回程票
3. aisle seats 靠走道的座位
4. Cancellation 取消

主旨

親愛的林女士：

我是iDigital Technical Enterprise的Becky。麻煩您儘可能地早一點預定兩張這個星期六早上到香港的來回機票。Lee總裁與我必須為巡迴展覽進行一趟商務旅行。一如往常地您已有我們的護照資料。

由於我們並不確定回來的確切時間，因此請讓機票保持為不定期的回程票。兩個座位都必須是靠走道的座位。我們兩個的座位沒有升級的必要。

請直接將電子機票直接以電子郵件寄到我的信箱: becky068@hotmail.com.

由於我們這一季會進行非常緊湊的行程，因此依據我們所簽屬的合約，必須提供可取消的制度。

獻上最好的祝福，

Chapter 2 Road Show Invitation（參展邀請函）

Scenario（情境）：
邀請工具機供應鏈管理伙伴共同參與參展

TO: SCM Partners

FROM: Becky Chen

DATE: 7 July 2016

SUBJECT: roadshow in Hong Kong

Invitation: Please join the roadshow with us

Dear valuable partners,

We warmly invite you to participate in the "Asia Machinery Roadshow" on 30th November 2016 in Hong Kong.

As it is the largest machinery roadshow in Asia, we deeply wish that you can join us since we are in the same Supply Chain Management. Demonstrably, this is the best opportunity for us to be present by the end of this year.

Please note the following details for your reference:

Date: 30th November 2016 (Wednesday)

Time: 8am to 5pm

Venue: Four Springs Hotel, Hong Kong

Please kindly find the attached agenda for your attention.

Thank you.

Yours sincerely,

Becky Chen

Senior Vice President

iDigital Technical Enterprise

No. 144, MinSheng Road, Taichung City, Taiwan, ROC

e-mail: hr@idigital.com.tw

Tel: +886-4-2218-1122

Fax: +886-4-2218-1123

Vocabulary or Phrases:

1. roadshow 巡迴展覽
2. warmly 熱烈地
3. demonstrably 可論證地、顯然地

主旨

邀請：請與我們一同參加巡迴展

親愛的尊貴夥伴們：

我們熱烈地邀請您於2016年11月30日來香港參加「亞洲機械設備巡迴展」。

身處相同供應鏈的管理部門，我們誠摯地希望您可以一同參加這個亞洲最大的機械設備巡迴展。顯然地，這是今年年底前我們能夠參與的最佳機會。

提供以下細節供您參閱：

日期：2016年11月30日星期三
時間：早上8點到下午5點
集合地點：Four Springs Hotel, Hong Kong

煩請您留意以下附件的議程表。
謝謝您。

您誠摯的朋友，

Chapter 3 Hotel Reservation（旅館預約函）

Scenario（情境）：
向旅館預訂房間

TO: Receptionist of Four Seasons Hotel

FROM: Becky Chen

DATE: August 7, 2015

SUBJECT: Reservation on August 23rd ~ 25th

Dear Manager,

Please kindly book double room single occupancy from 23rd to 25th August, 2016 in the names of Mr. Allen Hsu and Mrs. Mary Chou from Taipei. They have to attend the annual roadshow for our company.

Non-Smoking and high rise rooms will be the preference. In addition, please note that the wake-up call is 7:00 in the morning for the delegation.

Your bill in this regard may be sent to our headquarter office for settlement as usual.

Please send the notification message via e-mail to hr@idigital.com.tw by the end of tomorrow for our confirmation purpose.

Thank you.

Yours faithfully,

Becky Chen

Senior Vice President

iDigital Technical Enterprise

No. 144, MinSheng Road, Taichung City, Taiwan, ROC

e-mail: hr@idigital.com.tw

Tel: +886-4-2218-1122

Fax: +886-4-2218-1123

Vocabulary or Phrases:

1. receptionist 接待員
2. preference 偏好
3. delegation 代表團
4. settlement 結算

主旨

親愛的經理：

煩請在來自台北的Mr. Allen Hsu和Mrs. Mary Chou名下預定從2016年8月23日至8月25日雙人房。他們兩位需要參加本公司辦理的年度巡迴演出。

禁菸和高樓層將是挑選房間的偏好。另外，請註明為代表團叫醒服務的時間為早上7點。

您此次的帳單可能會一如往常地被送至總部的辦公室進行結算。

請最晚在明日之前藉由電子郵件寄出通知訊息至hr@digital.com.tw以利我們確認程序。

感謝您。

您忠誠的，

Chapter 4 Car Rental（租車函）

Scenario（情境）：
向租車公司預約車輛

Becky Chen

iDigital Technical Enterprise

513, Sec. 1, Taiwan Boulevard,

Taichung City, 43301 Taiwan

8-Aug-15

Joyce Lam

Rent-A-Car Enterprises

100 Main Street • Keeseville • NY • 12944

Phone 518-555-2345

Fax 518-555-2346

Dear Ms. Lam,

I found your information from local newspaper and I figure out that you are the reservationist of Rent-A-Car Enterprises in this area.

Due to the unavailability of the roaming service of my mobile phone carrier in this region, I have no choice but to send you this fax concerning the reservation of a 4 doors sedan for 3 days with unlimited mileages. The lease will be effective on 8:00 a.m. 13th August to 8:00 a.m. 16th if possible. Delivery service is a must since I am not able to access local transportation from my hotel.

Please confirm the above reservation with the quotation and the associate policies via sending me the fax to 518-555-1234 with attention to room #834, Ms. Chen.

Due to the tight schedules I have to confirm with my clients, your prompt reply will be highly appreciated.

Sincerely yours,

Becky Chen

Senior Vice President

iDigital Technical Enterprise

No. 144, MinSheng Road, Taichung City, Taiwan, ROC

e-mail: hr@idigital.com.tw

Tel: +886-4-2218-1122

Fax: +886-4-2218-1123

Vocabulary or Phrases：

1. reservationist 預約登記服務員
2. unavailability 無法取得
3. roaming 漫遊
4. have no choice but to 別無選擇
5. unlimited mileages 無限制里程數

主旨

親愛的Lam小姐：

我發現您在當地報紙上的訊息，並發現您是Rent-A-Car Enterprises在此地區的預約登記服務員。

由於我的手機通信商的關係，在此地區無法提供網路漫遊的服務，我別無他法，只能藉由傳真告知關於一份為期3天不限里程數的四門轎車訂單。如果可行，出租契約將會在8月13日早上8點至16日早上8點生效。交車服務是必須的，因為我沒有辦法從飯店使用當地的交通工具。

請確認上述的訂單，附上報價和相關規定，藉由傳真送至518-555-1234並註明#834房的Ms. Chen。

由於必須與客戶確認緊湊行程的關係，您的快速回覆將會是被高度讚賞的。

誠摯的，

Chapter 5 Transportation Delay（交通延遲信函）

Scenario（情境）：
因路上耽擱的致歉函

TO: Joyce Smith

FROM: Becky Chen

DATE: August 8, 15

SUBJECT: Urgent notification concerning the delay

Dear Joyce,

This is an urgent notification regarding the absence of the meeting tomorrow 8:00 a.m. due to the severe snowstorm condition in Denver International Airport.

The local weather bureau has just issued a severe snowstorm watch effectively immediately, which causes the airport to shut down unconditionally. Obviously, I am not capable of being physically present the meeting tomorrow.

There is no sign when the airport will be re-opened. I have to sincerely apologize for the unexpected weather condition. However, once the airport is open, you will be the first one I will contact straightaway.

Once again, I am awfully sorry for the delay of the meeting. Your understanding would be beyond description.

Becky Chen

Becky Chen

Senior Vice President

iDigital Technical Enterprise

No. 144, MinSheng Road, Taichung City, Taiwan, ROC

e-mail: hr@idigital.com.tw

Tel: +886-4-2218-1122

Fax: +886-4-2218-1123

Vocabulary or Phrases:

1. severe snowstorm watch 暴風雪警報
2. unconditionally 絕對地
3. unexpected 出乎意料地
4. straightaway 立即
5. beyond description 超出想像地

主旨

親愛的Joyce：

這是一件緊急通知，由於Denver國際機場的嚴重暴風雪，明日早上8點的會議本人將無法出席。

當地的氣象局剛剛已發出了立即生效的暴風雪警報，導致機場完全關閉。很明顯地，我無法在明日的會議中現身。

機場並未明確告知何時會重新開放。我必須為無法預測的天氣狀況致上真誠的歉意。一旦機場開放，您將會是第一位收到通知的。

本人再一次為會議的延期感到非常抱歉。您的包容和理解將會是難以形容的。

Chapter 6 Business Announcement（公司公告函）

Scenario（情境）：
公司將於年底進行土地開發案相關作業，特此公告

To: All the employees

From: Stephen Lu

Date: December 12

Subject: Concerning the company predetermined plan

Dear all,

We have decided to start a land development project at the end of this year. In order to execute the project on time, there are some prerequisites we already done: encompassing market assessment, acquire a potential land and apply for the development issues.

Furthermore, lots of companies and organizations had shown their strong interest to this development plan and decided to invest it recently. The following are the list of the new partners that we have gained by now, and also the money they invest for the plan.

- Official Land Investment Company US $200,000
- TAIWAN Land Progress Company US $150,000
- Land Cooperation group US $100,000

- TL develop Company US $80,000

PROSPECT:

Since the development plan has been done. Base on the economic development trend nowadays, the huge interests that bring to our company could be truly assured.

Sincerely,

Stephen Lu

Stephen Lu

J.ZON Land Enterprise

P.O. Box 123 • Keeseville • NY • 12944

Phone 518-555-2345

Vocabulary or Phrases:

1. project 計畫
2. prerequisites 先決條件
3. encompassing 包含、涵蓋
4. assessment 評估
5. acquire 獲得
6. assured 確定

主旨

親愛的各位：

公司決定在今年年底前開始執行一個土地開發計畫，為了讓計畫能夠如期進行，我們已經完成了一些事前的準備工作，包括整個市場的評估、取得有開發潛力的土地、土地開發計畫的相關申請作業。

此外，最近有許多公司和組織都對此開發計畫表達濃厚的興趣，並且決定進行投資。以下是我們公司目前所取得的新合作夥伴名單，以及他們投資該計畫的金額。

未來展望：
一旦開發計畫順利完成，以現今經濟發展的趨勢來看，可以預見此計畫在未來將會帶給我們的龐大利益。

誠摯的，

Chapter 7 Changes the Schedule（變更行程函）

Scenario（情境）：
行程計畫需要變更

From: Tim Duncan

To: Tony Parker

Fax number: 215-3-0507-1499

Oct.21 05:20 PM

PAGE 1 of 1

Dear Tony,

Thank you for sending the schedule for this week you've made for me.

Everything looks fine, expect for a few minor issues. First and forward, I'll be meeting up with some friends and having a dinner with them at a restaurant called Hell's Kitchen after the meeting tomorrow in the afternoon. Please kindly contact them and reserve a table for five at 7:00 p.m. Furthermore, I have to attend an important business conference on Wednesday. That means you have to make the adjustments as soon as possible, so I won't be absent from that meeting. There is just one more thing, on Friday afternoon, and I've promise my daughter not to be late for her School Sports Day. I hope that you can make me available for my family time on that day.

Thank you for the arrangements you have made so far. I hope that these changes do not create too much inconvenience to you.

Truly,

Tim

Vocabulary or Phrases:

1. minor 微小的
2. First and forward 首先
3. Furthermore 此外
4. absent from 缺席
5. inconvenience 不方便

主旨

親愛的Tony：

謝謝你寄給我這一週你替我安排的行程。

除了一點點小問題外，其他地方都很好。明天下午會議過後，我會和朋友見面並在地獄廚房吃晚餐。請你幫我和他們預定明天晚上7點的位子，總共有5個人。除此之外，星期三我必須參加一個重要的商業會議。這也代表需要請你盡快做出行程調整，這樣我才不會無法出席那場會議。還有一件事，星期五中午我已經答應我女兒會準時出席她們學校的運動會。我希望你可以讓我有時間與家人相處。

非常感謝你替我安排的行程。希望這些小變更不會造成你的困擾。

真摯的，

Chapter 8

Approval of a Business Case（商業案件同意函）

Scenario（情境）：

建設案送出，等待通過

To: Adam Levine

From: Steve Rogers

Date: January 1st

Subject: Hydroelectric Power Plant Building Plan

Attachment: Power Generating Evaluation

Dear Adam,

We appreciate your insight review. Please kindly find the enclosed power generating evaluation. Please let me know if there's' any question occurs.

LTA Power Company Hydroelectric Power Plant Building Plan

By Steven Rogers

Purpose:

For the shortage of the electricity in the Greater Taichung area, LTA power company have decided to build another hydroelectric power plant at the upstream of the river.

Construction Details:

The construction of the power plant will be taken over by the LTA construction company.

Construction cite: Around Chung-Shin Road.

Construction time: Approximately 20 years.

Power Generating Evaluation:

Name	Data
Water Intake	Wu River
Maximum Diversion	$41m^3$/sec
Diversion Tunnel Diameter	4.4m
Water Storage Capacity	748m
Effective Drop	1,100m
Power Generation	100,000KW

Best Wishes

Steven Rogers

Steven Rogers

Number:0936520143

Planning Office Manager

LTA power company

Vocabulary or Phrases：

1. appreciate 感激、欣賞
2. insight 有觀察力
3. evaluation 評估
4. occur 發生
5. hydroelectric power plant 水力發電廠
6. upstream 上游
7. intake 引水量
8. diversion 分水量
9. diameter 直徑

主旨

親愛的Adam：

我們欣賞您有洞察力的評論，敬請翻閱附上的發電評估，如果有碰到任何問題，請讓我知道。

LTA電力公司水力發電廠建造計畫

By Steven Rogers

目的：

為了因應大台中地區的電力短缺，LTA電力公司決定在烏溪的上游建造另一座水力發電廠。

建造詳細內容：

發電廠的建造將由LTA建設公司接手

建設地點：接近中清路

建設時間：大約20年

發電評估：

名稱	數據資料
引水處	烏溪
最大分水量	41立方公尺／秒
分水通道的直徑	4.4公尺
存水量	748公尺
有效發電高度	1,100公尺
電力生產	100,000千瓦

致上最好的祝福，

Customer Relationship Management
顧客關係管理信函

Module 06

Chapter 1 Creating New Customers（開發新客戶函）

Scenario（情境）：
房仲業欲開發新客戶

TO: Steven Chen

FROM: Becky Chen

DATE: 10-Jul-15

SUBJECT: Expecting your re-visit

Dear Mr. Chen,

As the marketing Manager of 21st Century, I would like to personally welcome you to visit our company again. My staff, Ms. Jane Liao will be your dedicated customer service representative to give you a detailed financial planning if you are interested in purchasing our new apartments in uptown Taipei.

In the past 20 years, 21st Century is the leading corporation in real estate arena in a row. We have created true happiness in life for tasted individuals. We have financial specialists to assist you for your mortgage planning. Dream house is no more unaffordable when 21st Century is physically present.

Quality is always number one issue for our products, which you can absolutely count on us. If you have furthermore questions, you can directly reach Ms. Jane Liao, whose mobile number is 0963-443-517. The service is available around the clock.

Once again, thank you for your visit to 21st Century.

Sincerely,

Becky Chen

Manger, Marketing Devision

iDigital Technical Enterprise

No. 144, MinSheng Road, Taichung City, Taiwan, ROC

e-mail: hr@idigital.com.tw

Tel: +886-4-2218-1122

Fax: +886-4-2218-1123

Vocabulary or Phrases:

1. dedicated 專屬的
2. representative 代表人員
3. in a row 連續
4. tasted individuals 有品味的人
5. mortgage 貸款

6. count on 依靠、相信

7. around the clock 全天候不休

主旨

親愛的Chen先生：

身為21st Century的行銷經理，我誠摯地邀請您再次來參觀本公司。如果您對於購買我們在台北住宅區的公寓有興趣，我的員工，Ms. Jane Liao將會是您專屬的客戶服務代表，並為您提供詳細的財務規劃。

在過去的20年，21st Century一直都是房地產競爭市場上的領導企業。我們為有品味的人創造了真正的快樂。我們有理財專家來協助您規劃貸款方案。只要21st Century還在，您夢想中的房子將不再是買不起的。

品質永遠是我們的產品所最關切的，而我們是絕對可靠的。如果您有任何詳細問題想詢問，您可以直接聯繫Ms. Jane Liao，行動電話為0963-443-517。24小時提供服務，不休息。

誠摯的，

Maintaining Old Customers （留住舊顧客函）

Scenario（情境）：
傢俱店欲邀舊顧客回店消費

TO: Louise Pan

FROM: Tom Lee

DATE: 8/5/15

SUBJECT: Special discount exclusively for you

Dear Mrs. Pan,

It has been a while that we did not see you in our store. We do understand that the tight schedules keep you busy as usual. We do expect you to come back to our store and we would like to offer you a special discount exclusively for you.

In the past 6 months, you have purchased a certain amount of furniture from us. It will be a great honor of us if you consider visiting us again by the end of next month. In order to show our sincerity, we will provide you 35% discount if you purchase our product by the end of the aforementioned date. In addition, we are more than happy to provide the door-to-door delivery service without further charges.

We are looking forwards to seeing you in our store in the near future.

Truly,

Tom Lee

Senior Manager of Customer Service

iFurnture Enterprise

No. 122, Peace Avenue, Taichung City, Taiwan, ROC

e-mail: crm@ifurniture.com.tw

Tel: +886-4-2218-1122

Fax: +886-4-2218-1123

Vocabulary or Phrases:

1. It has been a while 已經過了一段時間
2. Exclusively 特別地
3. door-to-door 完整的到府服務

主旨

親愛的Pan女士：

我們已經有一段時間沒有在店裡看到您的身影。我們明白您緊湊的行程讓您如往常般忙碌。我們期待您再來光顧我們的商店，我們將特別為您提供專屬的折扣。

在過去的6個月，您向我們購買了一些傢俱。如果您考慮在下個月月底前再次光臨本店，這將會是我們的榮幸。為了表示我們的誠意，如果您在上述日期之前至本店消費，我們將提供您35%的折扣。此外，我們很樂意提供免費送貨到府的服務。

我們期待您在不久的將來前來光顧。

真摯的，

Chapter 3 Customer Satisfaction（顧客滿意度信函）

Scenario（情境）：

進行顧客滿意度問卷調查

TO: HaiDen Cheng

FROM: Kevin Lai

DATE: 2015-08-05

SUBJECT: Customer Satisfaction Survey

Dear Mr. Cheng,

This is the Manager of customer satisfaction division of iDigital Technical Enterprise.

You purchased our coreless phone last week. Your satisfaction is our biggest concern. Therefore, we humbly ask you to do us a favor via visiting our web site to provide the feedback.

We deeply appreciate that if you could go to the following link and it will only take you a few more minutes to fill out the on-line questionnaires.

Http://www.crm.idigital.com.tw

After answering all the questions, you will be granted a secret gift from us within a week. Your prompt signing up to the above web site is highly recommended and appreciated.

Sincerely yours,

Kevin Lai

Manager, customer satisfaction division

No. 144, MinSheng Road, Taichung City, Taiwan, ROC

e-mail: kevin_lai@idigital.com.tw

Tel: +886-4-2218-1122

Fax: +886-4-2218-1123

Vocabulary or Phrases:

1. humbly 謙虛地
2. do us a favor 幫我們一個忙
3. questionnaires 問卷

主旨

親愛的Cheng先生：

我是iDigital技術企業的客戶滿意度調查部門的經理。

上週您購買了本公司的無線電話。您的滿意度是我們最關心的。因此，我們虛心地請您透過我們的網站提供回饋。

如果您可以點進下面的連結為我們填寫網路問卷，我們將會非常感謝，這只需要您幾分鐘的時間。

Http://www.crm.idigital.com.tw

當您回答完所有的問題之後，您將在一個星期內收到我們準備的一份神秘禮物。您對於上述網址的立即填寫，是我們強烈建議且非常感謝的。

您忠誠的，

Chapter 4

Customer Loyalty Pursuit（尋求顧客忠誠度信函）

Scenario（情境）：透過園藝服務尋求顧客的忠誠度

TO: Daniel Lo

FROM: Jenny Wang

DATE: 5-Aug-15

SUBJECT: Customer Loyalty Pursuit

Dear Daniel,

It's has been our greatest honor serving you concerning the gardening tasks in your esteemed house in the past 4 years. We mutually understand that you are our good customer and we are your excellent partner to make your gardening work easier and more fantastic.

As you know, there are plentiful gardening service providers around the city. Demonstrably, we have many loyal customers in the past years and we believe that you are one of them. For the purpose of strengthening our customer services in order to increase the competitiveness in this field, we have the toll-free number for you, 0800-765-354, with 7x24 services.

Whenever you need the gardening services, **iGardening Corporate** will definitely provide the best services and you can count on us.

Best,

Jenny Wang

Manager, customer service division

iGardening Corporate

No. 100, Main Avenue, Taichung City, Taiwan, ROC

e-mail: jenny@igerdening.com.tw

Tel: +886-4-2218-1122

Fax: +886-4-2218-1123

Vocabulary or Phrases:

1. Pursuit 追求
2. Mutually 互相地、彼此地
3. Plentiful 很多的、豐富的
4. Demonstrably 可被證實地
5. Competitiveness 競爭

主旨

親愛的Daniel：

在過去的四年裡，對於為貴住宅做園藝服務一直是我們最大的榮幸。我們彼此都知道，您是我們的好客戶，而我們也是您優秀的夥伴。透過我們的幫助，您的園藝工作能夠更輕鬆、更完善。

如您所知，周邊城市有許多的園藝服務供應商。可被證實地，在過去的幾年裡，我們有許多忠實客戶，且我們相信，您也是其中之一。為了加強我們的客戶服務，以增加在這一領域的競爭力，我們將提供您全年無休的免費電話服務，0800-765-354。

無論何時您有園藝服務的需要，您可以指望我們，iGardening企業肯定會為您提供最好的服務。

致上最佳祝福，

Chapter 5 Handling Customer Complaints （處理顧客抱怨信函）

Scenario（情境）：
及時處理客訴案件

TO: Peggy Fang

FROM: Bruce Lam

DATE: 2015-08-07

SUBJECT: An apology from **iDigital Technical Enterprise**

Dear Ms. Fang,

Firstly, on behalf of **iDigital Technical Enterprise**, I have to sincerely apologize to you regarding the complaint you filed last Monday. I feel extremely sorry that our customer service representative did not handle your request immediately and appropriately. We will definitely enhance our job training to next level.

Secondly, concerning the product you purchased, the 42-inch TV, we will replace that by the end of this week. Mr. Chen will contact with you momentarily respecting the delivery time that is convenient to you without any charges.

Thirdly, we always cherish the comments and feedbacks from our customers as the first priority. If you have any questions, please do not hesitate to call the service line which is 0800-234-876.

We are looking forwards to seeing you in our store in the near future.

Sincerely yours,

Bruce Lam

Vice President, Customer Service Division

iDigital Technical Enterprise

No. 144, MinSheng Road, Taichung City, Taiwan, ROC

e-mail: bruce_Lam@idigital.com.tw

Tel: +886-4-2218-1122

Fax: +886-4-2218-1123

Vocabulary or Phrases:

1. on behalf of 代表…
2. representative 代表人
3. appropriately 適當地
4. momentarily 立刻、隨時

主旨

親愛的Fang小姐：

首先，代表iDigital技術企業，關於您上週一提出的投訴，我們真誠地向您致歉。我感到非常抱歉，我們的客戶服務代表沒有立即且適當地處理您的請求。我們會確實加強我們的工作培訓，以期達到另一個水準。

其次，關於您所購買的42英寸電視，我們將在本週前為您做替換。陳先生將會立即與您聯繫，談論關於您方便的交貨時間，當然，這是免費提供的。

第三，我們永遠珍視客戶的意見和回饋，並將其視為第一優先。如果您有任何疑問，請不要猶豫，打我們的服務電話0800-234-876。

我們期待在不久的將來，能夠看到您再次光臨本店。

您忠誠的，

Chapter 6 Providing Technical and Management Supports（提供技術和經營幫助函）

Scenario（情境）：
新藥物合作和經營支援

PROPOSAL

From: BioChem Technology Company

To: Tiesto Products Company

Date: November 29

Subject: Technical and Management supports

BioChem Technology Company proposes to supply the following services to Tiesto Products Company for a monthly fee of US$7,000:

1. Assign three full-time, experienced bio-tech engineers to work with Tiesto's research laboratory in Taiwan with the goal of improving the doses and concentrations of the cancer curing products.
2. Equip the Tiesto representative with appropriate technical, managerial, and administrative supports.
3. Provide 400 sq. ft. of lab space for Tiesto's reliability engineering tests (additional space available).
4. Provide four men of lab work assistances during the research.
5. Furnish written reports at reasonable time intervals as required by Tiesto.

RESTRICTIONS AND NOTATIONS:

1. The BCTC laboratory is equipped with usual testing equipment, such as chart recorders, force and temperature measuring instruments, and common hand tools. The cost of special-purpose testing equipment is not included in these proposed fees.
2. Fees presented in this proposal include employee benefits and on-island travel and incidental expenses. Expenses for traveling abroad for research purpose will be billed to Tiesto separately.
3. BCTC proposes to provide services on a monthly contract basis, payable in advance each month. We also propose that the contract be cancelable by either party with one month's notice.
4. A set-up charge of US $4,500 will be required to cover expenses for BCTC management to survey each of Tiesto's vendors in Taiwan.

Sincerely yours,

James Bay

James Bay

Number: 0936520143

Planning Office Manager

BioChem Technology Company

Vocabulary or Phrases:

1. propose 提出
2. appropriate 適當的
3. managerial 管理的
4. instrument 儀器
5. incidental expense 臨時費用

主旨

BioChem科技公司提出了以每月7000美金的費用為Tiesto產品公司提供以下的服務。

1. 派出3名全職、有經驗的生物科技工程師協助Tiesto在台灣的研發實驗室中，修正癌症治療藥物的劑量與濃度。
2. 在Tiesto的代表席中給予適當的技術、管理、行政協助。
3. 提供400平方英尺的實驗空間來進行Tiesto的完成度工程測試（可另外添加空間）。
4. 在研究時，提供4人協助實驗。
5. 在合理的時間間隔中，提供Tiesto要求的書面報告。

須知規範：

1. BCTC實驗室設有一般的檢測工具，如：紀錄表、力測試儀、溫度計與一般的工具。特殊檢測儀器的費用並不包含在提出的費用中。
2. 提出的費用包含員工費用、國內旅行和臨時費用。以研究為目的的國外旅遊將會向Tiesto分開報帳。

3. BCTC提出以每月簽約的方式提供服務，每個月提前付費。我們也提出合約可以由任何一方在一個月前提前告知情況下可取消。
4. BCTC將向Tiesto收取調查Tiesto在台灣供應商的4,500美金的費用。

您忠誠的，

Chapter 7 Following Schedule Informed （告知部屬重要事項信函）

Scenario（情境）：
主管告知部屬交辦事項及行程

To: Candy Wu

From: Kevin Lai

Date: January 8th

Subject: Road show assistance

Dear Candy,

As you know that we are going to join the road show in Japan on February 12th. Undoubtedly, you are one of the best color analysts in our company. Hence, I am sincerely inviting you to go with us and expand your expertise. The company will reimburse your spending when you come back.

To increase our company's competitiveness, I'd like to urge all employees to dedicate themselves to our company's image operation, especially through the media. There will be workshops held in the City Hall from time to time in October. Everyone is encouraged to participate and provide the appropriate feedbacks to the company. Incentives will be applied if the contributions are huge.

We received tremendous orders from our collaborative partners. Next quarter will be a busy season for us. Stock options and bonus will be applied simultaneously to those people who fulfill outstanding achievements.

I personally really appreciate every single effort, passion, and enthusiasm you paid for our company. The bright future of the company will be definitely built based the team work.

Sincerely,

Kevin Lai

Vocabulary or Phrases:

1. undoubtedly 無疑地
2. reimburse 補償
3. dedicate 奉獻
4. encouraged 感到鼓舞
5. contributions 貢獻
6. simultaneously 同步
7. fulfill 履行、達成
8. achievements 成就
9. enthusiasm 熱情、熱忱

主旨

親愛的Candy：

如妳所知，我們2月12日要去日本參加巡迴展覽。無疑地，妳是公司裡最優秀的色彩分析師之一。因此，我誠摯地邀請妳和我們一起前往並拓展妳的專業知能。妳回來時，公司會付清此行的費用。

為了增加公司的競爭力，我希望所有員工都貢獻一己之力提升公司的企業形象，特別是透過媒體管道。市政府10月會不定時舉辦一些工作坊。鼓勵每個人去參加並提供一些合適的回饋給公司。若貢獻很多，公司會提供獎勵。

我們從合作公司那得到大量的訂單。下一季對我們來說將是旺季。股票與獎金將同步提供給那些達成優秀成就的人們。

我個人真的非常感謝你們每一個人為公司所付出的努力、熱情和熱忱。公司光明的未來，無疑是建立在團隊合作之上。

誠摯的，

Chapter 8 Budget of miscellaneous Items（降低雜項預算函）

Scenario（情境）：
公司欲降低雜項預算，並詢問員工的意見

To: All employees

From: Ann Stark

Date: December 10

Subject: Cutting budget

I've just reviewed our expenditures for the first quarter of the year. I'm happy to report to all of you that most categories were within our budget goals.

One area in which we overspent, however, was Miscellaneous Office Expenses. This category includes phones, faxes, shipping, postages, and so on. I would like to ask your help in reducing this area of expenses even more.

Here are a few suggestions:

1. Contact clients by e-mail as much as possible.
2. Notice the timing of making international calls. Always keep in mind that the most expensive time to make international calls is between

8:00 a.m. and 4:00 p.m.

3. Take advantage of wastepaper. It can be a good way to cut budgets of printing, or you can try to print by double-sided printing.
4. Please do not hesitate to contact HR department If you have good ideas to help our company cut budget in our routine operations.

Thank you for your cooperation and let's make the company more perspective.

Yours truly,

Ann

All-Cut electronic cooperation

Phone 2218-4598

Email ranky11322@gmail.com

Vocabulary or Phrases:

1. expenditure 支出、開銷
2. Miscellaneous Office Expenses 雜項支出
3. keep in mind 留意
4. take advantage of 好好利用
5. double-sided printing 雙面印刷
6. routine 規律的、常規的
7. perspective 透視的、有願景的

主旨

我剛剛回顧了上一季我們的開銷。我很開心地向各位報告，我們在大多數種類的款項支出都維持在預算範圍內。

然而，有一個項目超出了預算，那就是雜項支出。這個類別包括電話、傳真、航運、郵資等，希望各位能幫忙減低這方面的開銷。

這裡有些建議：

1. 儘量以e-mail聯繫客戶。
2. 注意撥打國際電話的時機。請特別留意，國際電話最貴的通話時間是在早上8點到下午4點之間。
3. 好好利用廢紙。這是一個減低印刷預算的好方法，或者你也可以試著使用雙面印刷。
4. 如果你有好的想法幫助公司節省預算，請不要猶豫、在例行性運作期間，聯絡我們人力資源部門。

感謝您的合作，讓我們一起為了公司創造遠景吧。

誠摯的祝福，

Chapter 9 Adjustment for one's Comment（意見調整告知函）

Scenario（情境）：

感謝對方意見，但客觀上無法配合

To: Dan Brown

From: Ann Stark

Date: March 14

Subject: An adjustment for your ideas

Dear Dan,

Thanks for your opinions respecting the school fair, and these comments are so precious that people would be definitely impressed by your great idea. Unfortunately, we don't get enough funding to carry out these plans. In this case, we have to find ways to raise more money. To be very honesty with you, under current situation, I have no choice but to put your brilliant ideas for the future.

I really appreciate your participation, and we hope that you can join our planning group if this is feasible to you. With your promising creativity, we can definitely worthy for this school fair in all aspects.

Sincerely,

Ann

Alice Stark

School Fair Planning Group

Phone 2218-3566

Email ranky11322@gmail.com

Vocabulary or Phrases:

1. respecting 關於
2. definitely 絕對地
3. carry out 執行
4. raise 募集
5. participation 參與
6. feasible 可行的
7. promising 傑出的、有爲的
8. aspects 方面

主旨

親愛的 Dan：

感謝您提供關於學校園遊會的意見，而這些意見是如此珍貴，人們必定會對此印象深刻。不幸的是，我們沒有足夠的資金去執行這些意見。在這種情況下，我們必須想盡各種辦法來募集更多資金。坦白說，在當前情勢下，我別無選擇，只能將你良好的意見當作未來參考。

我真的很感謝你的參與，如果可行的話，希望你能加入我們的規劃小組。有了你傑出的創造力，我們絕對能讓這次學校園遊會活動在各方面都圓滿成功。

誠摯的祝福，

Module

07

English Proverbs 英文諺語

Module 07

Chapter 1 English Proverb 1 英語諺語（A~B）

英文諺語的使用，在商用英文學習上也佔有重要的地位。因爲英文諺語是古代很多人智慧的累積，所以在學習商用英文寫作的時候，也可以適當地參考，以提升在商用英文方面的能力。

A

1. A barking dog never bites 會叫的狗不咬人
2. A bird in the hand is worth two in the bush 一鳥在手，勝過兩鳥在林
3. A cat may look at a king 人人平等
4. A change is as good as a rest 調換一下工作是很好的休息
5. A dog is a man's best friend 狗是我們最好的朋友
6. A drowning man will clutch at a straw 狗急跳牆
7. A fool and his money are soon parted 笨蛋難聚財
8. A friend in need is a friend indeed 患難見眞情
9. A golden key can open any door 有錢能使鬼推磨
10. A good beginning makes a good ending 好的開始是成功的一半
11. A good man is hard to find 好人難尋
12. A house divided against itself cannot stand 家庭內訌難維繫
13. A journey of a thousand miles begins with a single step 萬丈高樓平地起
14. A leopard cannot change its spots 江山易改，本性難移
15. A little knowledge is a dangerous thing 淺學誤人
16. A little learning is a dangerous thing 淺學誤人

17. A miss is as good as a mile 失之毫釐，差之千里
18. A new broom sweeps clean 新官上任三把火
19. A penny saved is a penny earned 積沙成塔
20. A person is known by the company he keeps 近朱者赤，近墨者黑
21. A picture paints a thousand words 一張圖表勝過千言萬語
22. A place for everything and everything in its place 適才適所
23. A poor workman always blames his tools 不會撐船怪河彎
24. A problem shared is a problem halved 衆人拾柴火焰高
25. A prophet is not recognized in his own land 外國的月亮比較圓
26. A rising tide lifts all boats 水漲船高
27. A rolling stone gathers no moss 滾石不生苔，轉業不聚財
28. A soft answer turneth away wrath 和言足以息怒
29. A stitch in time saves nine 適時處理，事半功倍
30. A watched pot never boils 心急水不沸
31. A woman's place is in the home 女子以家爲重心
32. A woman's work is never done 婦女的家事永遠做不完
33. A word to the wise is enough 智者不必多言
34. Absence makes the heart grow fonder 小別勝新婚
35. Absolute power corrupts absolutely 絕對權力，絕對腐敗
36. Accidents will happen. 天有不測風雲，人有旦夕禍福
37. Actions speak louder than words 坐而言不如起而行
38. Adversity makes strange bedfellows 身處逆境不擇友
39. After a storm comes a calm 否極泰來
40. All good things come to he who waits 機會只給有準備的人

41. All good things must come to an end 天下無不散的筵席
42. All is grist that comes to the mill 善於利用一切事物
43. All publicity is good publicity 任何宣傳都是好宣傳
44. All roads lead to Rome 條條大路通羅馬
45. All that glisters is not gold 閃耀的不一定是金子
46. All the world loves a lover 戀愛中的人惹人愛
47. All things come to those who wait 皇天不負苦心人
48. All things must pass 事情總會過去
49. All work and no play makes Jack a dull boy 只學不玩的孩子會變傻
50. All you need is love 愛正是你需要的
51. All's fair in love and war 情場如戰場
52. All's well that ends well 結局好就好
53. A miss is as good as a mile 失之毫釐，差之千里
54. An apple a day keeps the doctor away 一天一蘋果，醫生遠離我
55. An Englishman's home is his castle 家是人的城堡
56. An eye for an eye, a tooth for a tooth 以牙還牙，以眼還眼
57. An ounce of prevention is worth a pound of cure 預防勝於治療
58. Another day, another dollar 做一天和尚撞一天鐘
59. Any port in a storm 慌不擇路
60. April showers bring forth May flowers 凡事必有因果關係
61. As thick as thieves 親密無間
62. As you make your bed, so you must lie upon it 自作自受
63. As you sow so shall you reap 種瓜得瓜，種豆得豆

64. Ashes to ashes dust to dust 塵歸塵，土歸土

65. Ask a silly question and you'll get a silly answer 發出愚蠢的問題，不值得得到適切的答案

66. Ask no questions and hear no lies 少問就不容易聽到假話

67. Attack is the best form of defence 攻擊是最好的防衛

B

1. Bad money drives out good 劣幣驅逐良幣
2. Bad news travels fast 好事不出門，壞事傳千里
3. Barking dogs seldom bite 會吠的狗不咬人
4. Be careful what you wish for 別輕易許願
5. Beauty is in the eye of the beholder 情人眼裡出西施
6. Beauty is only skin deep 美麗是膚淺的
7. Beggars should not be choosers 弱者無權
8. Behind every great man there's a great woman 偉大的男人背後總有一名偉大的女性
9. Better late than never 亡羊補牢，爲時未晚
10. Better safe than sorry 預防勝於後悔
11. Better the Devil you know than the Devil you don't 未知的敵人最危險
12. Better to light a candle than to curse the darkness 與其抱怨不如行動
13. Better to remain silent and be thought a fool that to speak and remove all doubt 沉默是金
14. Beware of Greeks bearing gifts 小心敵人笑裡藏刀

15. Big fish eat little fish 適者生存
16. Birds of a feather flock together 物以類聚
17. Blessed are the peacemakers 祝福生和平
18. Blood is thicker than water 血濃於水
19. Boys will be boys 小孩畢竟是小孩
20. Brevity is the soul of wit 言以簡爲貴
21. Business before pleasure 利重於義

Chapter 2 English Proverb 2 英語諺語（C~G）

C

1. Caesar's wife must be above suspicion 權貴之言難質疑
2. Charity begins at home 仁愛始於家
3. Charity covers a multitude of sins 施捨遮百惡
4. Cheaters never win and winners never cheat 騙子是絕不會成功的，而成功的人是絕不會騙人的
5. Children and fools tell the truth 孩子和傻子不說謊
6. Children should be seen and not heard 大人在講話，小孩別插嘴
7. Cleanliness is next to godliness 整潔近於美德
8. Clothes make the man 人要衣裝，佛要金裝
9. Cold hands, warm heart 刀子嘴，豆腐心
10. Count your blessings 對你擁有的一切感到慶幸和知足
11. Cowards may die many times before their death 勇者無懼
12. Cut your coat to suit your cloth 量入爲出

D

1. Dead men tell no tales 殺人滅口
2. Distance lends enchantment to the view 距離讓景色增添魅力
3. Do as I say, not as I do 別管我做得怎樣，照我說的去做
4. Do as you would be done by 己所不欲勿施於人
5. Do unto others as you would have them do to you 己所不欲勿施於人
6. Don't bite the hand that feeds you 勿恩將仇報

7. Don't burn your bridges behind you 不要自斷後路
8. Don't cast your pearls before swine 對牛彈琴
9. Don't change horses in midstream 不要中途變卦
10. Don't count your chickens before they are hatched 未到安全地，高呼犯大忌
11. Don't cross the bridge till you come to it 船到橋頭自然直，勿杞人憂天
12. Don't get mad, get even 不要生氣，要報復
13. Don't keep a dog and bark yourself 勿養鼠爲患
14. Don't leave your manners on the doorstep 不要把你的禮儀規矩留在家門外
15. Don't let the grass grow under your feet 馬上行動，勿拖延
16. Don't meet troubles half-way 別自尋煩惱
17. Don't mix business with pleasure 不要把正事和娛樂混在一起
18. Don't put all your eggs in one basket 不要孤注一擲
19. Don't put the cart before the horse 不要本末倒置
20. Don't rock the boat 不要打破局勢平衡
21. Don't shoot the messenger 不要遷怒於報信的人
22. Don't sweat the small stuff 別爲小事發愁
23. Don't throw pearls to swine 不要對牛彈琴
24. Don't teach your Grandma to suck eggs 別班門弄斧
25. Don't trust anyone over thirty 別聽信30歲以上之人的話（對方可能倚老賣老或雙方有代溝）
26. Don't try to run before you can walk 凡事按部就班

27. Don't try to walk before you can crawl 凡事按部就班
28. Don't upset the apple-cart 別打亂計畫，破壞任務
29. Don't wash your dirty linen in public 家醜不可外揚
30. Doubt is the beginning not the end of wisdom 懷疑，是智慧的萌芽，而不是結束

E

1. Early to bed and early to rise, makes a man healthy, wealthy and wise 早睡早起，使你身體健康、富裕和聰明
2. East is east, and west is west 東西文化大不同
3. East, west, home's best 金窩、銀窩，不如自己的狗窩
4. Easy come, easy go 來的快，去的也快
5. Eat, drink and be merry, for tomorrow we die 把握當下，及時行樂
6. Empty vessels make the most noise 沒有內涵的人才會炫耀
7. Enough is as good as a feast 足食猶如筵席
8. Enough is enough 適可而止
9. Even a worm will turn 狗急跳牆
10. Every cloud has a silver lining 雨過總會天晴
11. Every dog has its day 每個人都有出頭天的一天
12. Every Jack has his Jill 有情人終成眷屬
13. Every little helps 勿以善小而不爲（聚沙成塔；積少成多）
14. Every man for himself, and the Devil take the hindmost 人不爲己，天誅地滅
15. Every man has his price 有錢能使鬼推磨

16. Every picture tells a story 每張照片背後都有一個故事
17. Every stick has two ends 凡事總是有利也有弊
18. Everyone wants to go to heaven but nobody wants to die 每個人都想進天堂，卻沒有人願意面對死亡
19. Everything comes to him who waits 皇天不負苦心人

F

1. Failing to plan is planning to fail 無計畫就意味著失敗
2. Faint heart never won fair lady 懦夫難贏美人心
3. Fair exchange is no robbery 合理交換即無奪取
4. Faith will move mountains 精誠所至，金石為開
5. Familiarity breeds contempt 親不尊，熟生蔑
6. Fight fire with fire 以火攻火；以毒攻毒
7. Fight the good fight 謀定而後動
8. Finders keepers, losers weepers 誰撿到歸誰
9. Fine words butter no parsnips 花言巧語是無用的
10. First come, first served 先來先得
11. First impressions are the most lasting 第一印象記憶最深刻持久
12. First things first 最重要的事要先做
13. Fish always stink from the head down 上樑不正下樑歪
14. Flattery will get you nowhere 諂媚是不會使你得到任何好處的
15. Fools rush in where angels fear to tread 初生之犢不畏虎
16. For everything there is a season 萬物有時節
17. Forewarned is forearmed 防患於未然，有備無患
18. Forgive and forget 得饒人處且饒人

19. Fortune favors the brave 天佑勇者

20. From the sublime to the ridiculous is only one step 可敬與可笑之間只有一步之差

G

1. Genius is an infinite capacity for taking pains 天賦乃辛勞工作的無限能力

2. Genius is one percent inspiration, ninety-nine percent perspiration 天才是一分的靈感和九十九分的汗水

3. Give a dog a bad name and hang him 人言可畏

4. Give a man enough rope and he will hang himself 多行不義必自斃

5. Give credit where credit is due 對於值得讚揚之處要給予讚揚

6. Go the extra mile 多付出一點點；比別人多做一點

7. God helps those who help themselves 天助自助者

8. Good fences make good neighbours 君子之交淡如水

9. Good talk saves the food 妙語可餐

10. Good things come to those who wait 好事多磨

11. Great minds think alike 英雄所見略同

Chapter 3 English Proverb 3 英語諺語（H~L）

H

1. Half a loaf is better than no bread 有總比沒有好
2. Handsome is as handsome does 行爲美才是眞美
3. Hard work never did anyone any harm 努力工作對任何人都無害處
4. Haste makes waste 欲速則不達
5. He that goes a-borrowing, goes a-sorrowing 借債是不幸的開始
6. He who fights and runs away, may live to fight another day 好漢不吃眼前虧
7. He who hesitates is lost 猶豫者多失
8. He who laughs last laughs longest 鹿死誰手，尙未可知
9. He who lives by the sword shall die by the sword 玩火自焚
10. He who pays the piper calls the tune 出錢的人有權決定錢怎麼花
11. Hell hath no fury like a woman scorned 最毒婦人心
12. Hindsight is always twenty-twenty 事後諸葛
13. History repeats itself 歷史會重演
14. Home is where the heart is 心安即是家
15. Honesty is the best policy 誠實爲上策
16. Hope springs eternal 希望常在
17. Horses for courses 知人善任

I

1. If anything can go wrong, it will 事情如果還能更糟的話，它一定

會更糟

2. If a job is worth doing it is worth doing well 如果一件事值得去做，那麼就值得你去做好
3. If at first you don't succeed try, try and try again 屢敗屢戰
4. If God had meant us to fly he'd have given us wings 如果上帝想讓我們飛行，祂會賜予我們一對翅膀
5. If life deals you lemons, make lemonade 把吃苦當作吃補
6. If the cap fits, wear it 如果是眞的，就承認吧
7. If the mountain won't come to Mohammed, then Mohammed must go to the mountain 如果事情不能向你而來，你應該自己走向他
8. If the shoe fits, wear it 如果是眞的，就承認吧
9. If wishes were horses, beggars would ride 如果許願就能讓夢想成眞，包括乞丐在內的任何人也能擁有任何想要的東西
10. If you build it they will come 只要堅持就能達成目標
11. If you can't be good, be careful 可以玩得盡興，但不可任性胡來
12. If you can't beat them, join them 如果打不過他們，就投靠他們吧
13. If you can't stand the heat get out of the kitchen 如果再抱怨，那就別做了
14. If you lie down with dogs, you will get up with fleas 近朱者赤，近墨者黑
15. If you pay peanuts, you get monkeys 微薪養蠢材
16. If you want a thing done well, do it yourself 欲事成當自爲之
17. Ignorance is bliss 無知便是福

18. Imitation is the sincerest form of flattery 模仿是最眞誠的恭維
19. In for a penny, in for a pound 一不做，二不休
20. In the kingdom of the blind the one eyed man is king 蜀中無大將，廖化作先鋒；山中無老虎，猴子當大王
21. In the midst of life we are in death 我們的生活正在消失
22. Into every life a little rain must fall 每個人都會經歷波折
23. It goes without saying 不言而喻
24. It is best to be on the safe side 善加準備以防萬一總是較好
25. It is better to give than to receive 施比受更有福
26. It is easy to be wise after the event 不經一事，不長一智
27. It never rains but it pours 不鳴則已一鳴驚人
28. It takes a thief to catch a thief 以毒攻毒
29. It takes all sorts to make a world 屋漏偏逢連夜雨
30. It takes one to know one 知己知彼
31. It takes two to tango 孤掌難鳴
32. It's all grist to the mill 多多益善
33. It's an ill wind that blows no one any good 不管情況多麼壞，仍有人會得到好處
34. It's better to have loved and lost than never to have loved at all 不在乎天長地久，只在乎曾經擁有
35. It's better to light a candle than curse the darkness 與其咒罵一切，不如燃起希望之燈
36. It's better to travel hopefully than to arrive 懷著希望去旅行，比抵達目的地更愉快

37. It's never too late 永遠不嫌遲

38. It's never too old to learn 活到老學到老

39. It's no use crying over spilt milk 覆水難收

40. It's no use locking the stable door after the horse has bolted 為時已晚

41. It's the early bird that catches the worm 早起的鳥兒有蟲吃

42. It's the empty can that makes the most noise 半瓶水響叮噹

43. It's the singer not the song 是歌手唱得好，而不是歌曲本身

44. It's the squeaky wheel that gets the grease 會吵的小孩有糖吃

J

1. Jack of all trades, master of none 一個人什麼都會，但是什麼都不專精

2. Judge not, that ye be not judged 不要評斷人，免得你們被評斷

K

1. Keep your chin up 別灰心

2. Keep your friends close and your enemies closer 世界上沒有永遠的朋友，更沒有永遠的敵人

3. Keep your powder dry 做好一切準備

L

1. Laugh and the world laughs with you, weep and you weep alone 世界能與你同樂，但不能陪你哭泣

2. Laughter is the best medicine 開懷大笑是一劑良藥

3. Least said, soonest mended 多說誤事

4. Less is more 簡約見精華

5. Let bygones be bygones 不咎既往
6. Let not the sun go down on your wrath 不要含怒到隔天
7. Let sleeping dogs lie 不要之招惹不必要的麻煩
8. Let the punishment fit the crime 量刑判罪
9. Let well alone 得過且過，見好就收
10. Life begins at forty 四十不惑
11. Life is just a bowl of cherries 一切都十全十美
12. Life is what you make it 人生由你打造
13. Lifc's not all beer and skittles 生活並非飲酒作樂而已
14. Lightning never strikes twice in the same place 事不過二
15. Like father, like son 有其父必有其子
16. Little pitchers have big ears 隔牆有耳
17. Little strokes fell great oaks 滴水穿石
18. Little things please little minds 胸無大志，事事稱心
19. Live for today for tomorrow never comes 盡享現在的生活
20. Look before you leap 三思而後行
21. Love of money is the root of all evil 金錢是萬惡之源
22. Love is blind 愛情是盲目的
23. Love makes the world go round 愛使世界運轉
24. Love thy neighbour as thyself 千金難買好鄰居
25. Love will find a way 愛定出明路

Chapter 4 English Proverb 4 英語諺語（M~S）

M

1. Make hay while the sun shines 打鐵趁熱
2. Make love not war 創造愛，而非戰爭
3. Man does not live by bread alone 人不能只靠物質生存
4. Many a mickle makes a muckle 積少成多，集腋成裘
5. Many a true word is spoken in jest 戲言寓眞理
6. Many are called but few are chosen 蒙召喚的人多，而蒙揀選的人少
7. Many hands make light work 人多好辦事
8. March comes in like a lion, and goes out like a lamb 三月春天乍暖還寒
9. March winds and April showers bring forth May flowers 你不知道明天將會帶來什麼，要把握住今天
10. Marriages are made in heaven 天作之合
11. Marry in haste, repent at leisure 婚前不睜眼，婚後傻了眼
12. Might is right 強權即公理
13. Mighty oaks from little acorns grow 萬丈高樓平地起
14. Misery loves company 同病相憐
15. Moderation in all things 凡事適可而止
16. Money doesn't grow on trees 金錢來之不易
17. Money is the root of all evil 錢是萬惡之源
18. Money isn't everything 金錢非萬能

19. Money makes the world go round 有錢好辦事
20. Money talks 有錢能使鬼推磨
21. More haste, less speed 欲速則不達

N

1. Necessity is the mother of invention 需求是發明之母
2. Needs must when the devil drives 情勢所迫
3. Never give a sucker an even break 好人受欺
4. Never go to bed on an argument 不要帶著怒氣上床
5. Never judge a book by its cover 人不可貌相
6. Never let the sun go down on your anger 不可含怒到日落
7. Never look a gift horse in the mouth 好心當成驢肝肺
8. Never put off until tomorrow what you can do today 今日事，今日畢
9. Never speak ill of the dead 別說死者的壞話
10. Never tell tales out of school 家醜不可外揚
11. Nine tailors make a man 人要衣裝，佛要金裝
12. No man can serve two masters 人不能事二君
13. No man is an island 沒有人是一座孤島
14. No names, no pack-drill 謹慎就不會受到懲罰
15. No news is good news 沒有消息就是好消息
16. No one can make you feel inferior without your consent 如果你自己看得起自己，沒有任何人可以瞧不起你
17. No pains, no gains 不勞則無獲
18. Nothing new under the sun 太陽底下沒有新鮮事

19. Nothing is certain but death and taxes 人生在世死難逃，偷稅漏稅辦不到
20. Nothing succeeds like success 一有所成，成就自然接踵而來
21. Nothing venture, nothing gain 不入虎穴，焉得虎子

O

1. Oil and water don't mix 水火不容
2. Old soldiers never die, they just fade away 老兵不死，只是凋零
3. Once a thief, always a thief 偷盜一次，做賊一世
4. Once bitten, twice shy 一朝被蛇咬，十年怕草繩
5. One good turn deserves another 好心有好報
6. One half of the world does not know how the other half lives 貧富不相知
7. One hand washes the other 狼狽爲奸
8. One man's meat is another man's poison 各有所愛
9. One law for the rich and another for the poor 法不動大戶，刑不上大夫
10. One swallow does not make a summer 講禮貌不吃虧
11. One volunteer is worth ten pressed men 一人自願做，比兩個人被迫做還好
12. One year's seeding makes seven years weeding 一年草結子，七年除不盡
13. Opportunity never knocks twice at any man's door 機會難得
14. Out of sight, out of mind 眼不見爲淨

P

1. Patience is a virtue 忍耐是一種美德
2. Pearls of wisdom 名言警句
3. Penny wise and pound foolish 小事聰明，大事糊塗
4. People who live in glass houses shouldn't throw stones 投鼠忌器
5. Possession is nine points of the law 現實佔有，九勝一敗磨
6. Power corrupts; absolute power corrupts absolutely 絕對的權力造成絕對的腐化
7. Practice makes perfect 熟能生巧
8. Practice what you preach 以身作則
9. Prevention is better than cure 預防勝於治療
10. Pride goes before a fall 驕者必敗
11. Procrastination is the thief of time 延宕是時間之賊
12. Put your best foot forward 表現自己最好的一面

R

1. Revenge is a dish best served cold 君子報仇，十年不晚
2. Revenge is sweet 甜蜜的復仇
3. Rob Peter to pay Paul 挖東牆補西牆
4. Rome wasn't built in a day 羅馬不是一天造成的

S

1. See no evil, hear no evil, speak no evil 非禮勿視，非禮勿聽，非禮勿言
2. Seeing is believing 眼見爲憑
3. Set a thief to catch a thief 以毒攻毒

4. Share and share alike 平均分配
5. Shrouds have no pockets 生不帶來，死不帶去
6. Silence is golden, Speech is Silver 沉默是金，辯才是銀
7. Slow but sure 天網恢恢，疏而不漏
8. Spare the rod and spoil the child 不打不成器
9. Speak as you find 有幾分證據，說幾分話
10. Sticks and stones may break my bones, but words will never hurt me 棍棒石頭可能打斷我的骨頭，但話語決不會傷害我
11. Still waters run deep 大智若愚
12. Strike while the iron is hot 打鐵趁熱
13. Stupid is as stupid does 蠢人做蠢事
14. Success has many fathers, while failure is an orphan 無人願意承擔責任

Chapter 5 English Proverb 5 英語諺語（T~Y）

T

1. Take care of the pence and the pounds will take care of themselves 小事留意，大事順利
2. Talk is cheap 光說不練
3. Talk of the Devil, and he is bound to appear 說曹操，曹操到
4. Tell the truth and shame the Devil 摒棄一切顧慮據實直言
5. The apple never falls far from the tree 有其父（母）必有其子（女）
6. The best defense is a good offence 進攻是最好的防守
7. The best things in life are free 自由可貴
8. The bigger they are, the harder they fall 爬得越高，摔得越重
9. The bottom line is the bottom line 逾期不候
10. The boy is father to the man 江山易改，本性難移
11. The bread always falls buttered side down 屋漏偏逢連夜雨
12. The child is the father of the man 江山易改，本性難移
13. The course of true love never did run smooth 真愛無坦途
14. The customer is always right 顧客至上
15. The darkest hour is just before the dawn 黑暗即將過去，黎明就在眼前
16. The devil finds work for idle hands to do 小人閑居做歹事
17. The devil looks after his own 魔鬼照顧自家人
18. The early bird catches the worm 早起的鳥兒有蟲吃

19. The end justifies the means 為達目的，不擇手段
20. The exception which proves the rule 例外證明了定律的存在
21. The apple doesn't fall far from the tree 蘋果會掉在距離蘋果樹不遠處
22. The good die young 好人不長命
23. The grass is always greener on the other side of the fence 外國的月亮比較圓
24. The hand that rocks the cradle rules the world 母親影響孩子的未來，還掌握了世界的未來
25. The husband is always the last to know 丈夫永遠是最晚知道的
26. The leopard does not change his spots 本性難移
27. The longest journey starts with a single step 千里之行，始於足下
28. The more the merrier 多多益善
29. The more things change, the more they stay the same 事物不斷改變，本質卻永遠不變
30. The pen is mightier than sword 文勝於武
31. The price of liberty is eternal vigilance 自由的代價是永遠的戒備
32. The shoemaker's son always goes barefoot 有技藝或知識的人，常忽略了身邊最親近的人
33. The squeaking wheel gets the grease 會吵的小孩有糖吃
34. The truth will out 眞相總會水落石出
35. The wages of sin is death 罪惡的代價是死亡
36. The whole is greater than the sum of the parts 團結力量大
37. There are two sides to every question 事情本是一體兩面

38. There’s a time and a place for everything 什麼場合時間做什麼事
39. There’s an exception to every rule 世事無絕對
40. There’s always more fish in the sea 天涯何處無芳草
41. There’s many a good tune played on an old fiddle 寶刀未老
42. There’s more than one way to skin a cat 處理事情的方法不只一種
43. There’s no accounting for tastes 各有所好
44. There’s no fool like an old fool 老了還愚昧就是眞正的愚昧了
45. There’s no place like home 金窩、銀窩，都比不上自己的狗窩
46. There’s no smoke without fire 事出必有因
47. There’s no such thing as a free lunch 天下沒有白吃的午餐
48. There’s no such thing as bad publicity 出名不分好壞
49. There’s no time like the present 擇日不如撞日
50. There’s none so deaf as those who will not hear 眞正的聾子不是聽不見的人，是聽到卻選擇無視的人
51. There’s one born every minute 每分鐘都有笨蛋出生
52. There’s safety in numbers 人多勢衆
53. They that sow the wind, shall reap the whirlwind 惡有惡報
54. Third time lucky 前兩次失敗，第三次比較容易成功
55. Those who do not learn from history are doomed to repeat it 不向歷史學習，會重複歷史的失敗
56. Those who live in glass houses shouldn’t throw stones 不要五十步笑百步
57. Those who sleep with dogs will rise with fleas 近墨者黑
58. Time and tide wait for no man 時光不等人

59. Time flies 時光飛逝

60. Time is a great healer 時間會帶走傷痛

61. Time is money 時間就是金錢

62. Time will tell 時間會證明一切

63. To err is human; to forgive divine 人都會犯錯，會原諒才是聖人

64. To everything there is a season 萬物皆有定時

65. To the victor go the spoils 勝者爲王

66. To travel hopefully is a better thing than to arrive 理想總比現實好

67. Tomorrow is another day 明天會更好

68. Tomorrow never comes 一味等待明日，萬事成蹉跎

69. Too many cooks spoil the broth 人多手雜

70. Truth is stranger than fiction 事實總是比故事更離奇

71. Truth will out 眞相總會水落石出

72. Two blacks don't make a white 負負不等於正

73. Two heads are better than one 合作力量大

74. Two is company, but three's a crowd 人多嘴雜

75. Two wrongs don't make a right 負負不等於正

V

1. Variety is the spice of life 見識多使得生活更有趣

2. Virtue is its own reward 美德應發自內心

W

1. Walls have ears 隔牆有耳

2. Waste not want not 勤儉爲美德

3. What can't be cured must be endured 無法改變，便學會忍耐

4. What goes up must come down 風水輪流轉
5. What you lose on the swings you gain on the roundabouts 利弊皆有
6. What's sauce for the goose is sauce for the gander 對別人如何對自己也應如何
7. When in Rome, do as the Romans do 入境隨俗
8. When the cat's away the mice will play 山中無老虎，猴子稱大王；蜀中無大將，廖化當先鋒
9. When the going gets tough, the tough get going 當你開始垂頭喪氣時，將更加沮喪
10. What the eye doesn't see, the heart doesn't grieve over 眼不見，心不悲
11. Where there's a will there's a way 有志者事竟成
12. Where there's muck there's brass 賺錢的工作有人做
13. While there's life there's hope 有生命就有希望
14. Whom the Gods love die young 有才之人總短命
15. Why keep a dog and bark yourself? 越俎代庖（執行自己下屬該做之事）
16. Women and children first 女人和小孩優先
17. Wonders will never cease 奇事總是不斷發生
18. Work expands so as to fill the time available 擴大工作量以在規定時間內完成
19. Worrying never did anyone any good 杞人憂天

Y

1. You are what you eat 人如其食
2. You can choose your friends but you can't choose your family 有些事情可以選擇，有些則不可
3. You can't have too much of a good thing 物極必反
4. You can lead a horse to water, but you can't make it drink 師父引進門，修行在個人
5. You can't have your cake and eat it 魚與熊掌不可兼得
6. You can't make a silk purse from a sow's ear 不好的材料做不出好東西
7. You can't make an omelette without breaking eggs 有所成就，必須有所犧牲
8. You can't make bricks without straw 必要的要素缺一不可
9. You can't run with the hare and hunt with the hounds 魚與熊掌不可兼得
10. You can't teach an old dog new tricks 不必費神教老狗玩新把戲
11. You can't judge a book by its cover 人不可貌相
12. You can't win them all 有得有失
13. You catch more flies with honey than with vinegar 花蜜總比醋容易引誘他人
14. You pay your money and you takes your choice 每個人都有自己的選擇
15. You reap what you sow 一分耕耘，一分收穫
16. Youth is wasted on the young 人生中美好的青春時光，卻被不成熟、輕狂的自己，給浪費掉了

[illegible]

[illegible]

[illegible]

You can't have your cake and eat it.

[illegible]

[illegible]

[illegible]

[illegible]

[illegible]

11. You … your … you … your choice …

15. You reap what you sow.

16. Youth is wasted on the young.

Eyes Catching Business Slogans
吸睛的商業標語

Module 08

Business Slogan (Part 1) 企業標語1

Slogan是企業標語，也是該組織企業透過群體腦力激盪的結果，足以眞正表達該企業的經營理念，用非常洗練、精簡的字眼，來闡述該企業的形象，Slogan是非常適合商用英文寫作的參考，值得用心體會。

1. SONY: make.believe

SONY
make.believe

Source: http://sonyfanatic.com/2014/01/sony-drops-make-believe-slogan/

2. LG: Life's good

Source: http://www.lg.com/global/about-lg/corporate-information/at-a-glance/our-brand

3. Acer: explore beyond limits

Source: http://www.acer-group.com/public/index.htm

4. Canon: Delighting You Always

Delighting You Always

Source: http://www.canon.com.my/personal/web/company/about/corporate_identity

5. Panasonic: Better Life, A Better World

Panasonic

A Better Life, A Better World

Source: http://panasonic.net/brand/

6. Apple: Think Different

Source: https://www.apple.com/

7. Nokia: Connecting People

Source: http://company.nokia.com/en

8. htc: quietly brilliant

Source: http://www.htc.com/tw/

9. Asus: Inspiring Innovation · Persistent Perfection

Source: http://www.asus.com/us/

10. Motorola: intelligence everywhere

Source: http://www.motorola.com/us/home

11. Titus: Time is love

Source: http://www.solvil-et-titus.com/index.html

12. Rolex: The Crown of Achievement

Source : http://www.rolex.com/zh-hant

13. Casio: the unexpected extra

Source : http://www.casio-intl.com/asia-mea/en/

14. Movado: The art of time

Source : http://www.intl.movado.com/

15. Swatch: Time is what you make of it

swatch

Source : http://www.swatch.com/us_en/press.html

16. OMEGA: The sign of excellence

Source: http://www.omegawatches.com/

17. SEIKO: It's your watch that tells most about who you are

Source: http://www.seikowatches.com/home.html

18. Maxima: Because times are changing

Source: http://maximawatches.com/home.aspx

19. Accurist: No ordinary old timer

Accurist

EST. 1946

LONDON

Source: http://www.accurist.co.uk/

20. IWC: Since 1868. And for as long as there are men.

Source: http://www.iwc.com/en-us/

Business Slogan (Part 2) 企業標語2

1. Uni-President: Three Good and One Fairness

Source: http://www.uni-president.com/

2. Coca-Cola: Life Begins Here

Source: http://us.coca-cola.com/home/

3. Nestle: Good Food, Good Life

Source: http://www.nestle.com/

4. Kraft Foods: Make today delicious

Source : http://www.kraftrecipes.com/home.aspx

5. MARS: Pleasure you can't measure

Source : http://www.mars.com

6. Lotte: The better way of life

Source : http://global.lotte.com

7. Kellogg: The best to you each morning

Source: http://www.kelloggs.com

8. PepsiCo: Performance with Purpose

PEPSICO

Source: http://www.PepsiCo.com

9. Procter & Gamble (P&G):

Source: http:// www.pg.com

10. I-MEI: Good food provider

Source: http://www.imeifoods.com.tw/English/index_2.html

11. NVIDIA: The Way It's Meant to Be Played

Source: http://www.nvidia.com.tw/object/visual-computing-tw.html

12. EA GAMES: Challenge Everything

Source: http://www.ea.com/

13. CRYTEK: Envision. Enable. Achieve.

Source: http://www.crytek.com/company

14. SEGA: Welcome to the next level

Source: http://www.sega.com

15. Nintendo: Now you're playing with power

Source: http://www.nintendo.com/

16. AMD: The Future Is Fusion

Source: http://www.amd.com

17. MSI: Innovation with Style

Source: http://www.msi.com

18. KFC: It's finger lickin' good!

Source: http://www.kfc.com

19. Google: Don't be evil

Google

Source: https://www.google.com

20. Dixons: Brining life to technology

Source : http://www.dixonsretail.com/

Business Slogan (Part 3) 企業標語3

1. BMW: The ultimate driving machine

Source: http://www.bmw.com

2. FORD: Go Further

Source: http://www.ford.com/

3. Chevrolet: FIND NEW ROADS

Source: http://www.chevrolet.com/

4. Porsche: There is no substitute.

Source: http://www.porsche.com

5. Volvo: For life

Source: http://www.volvo.com

6. BENZ: The best or nothing.

Mercedes-Benz

Source: https://www.mercedes-benz.com

7. HONDA: The Power of Dreams

HONDA
The Power of Dreams

Source: http://www.honda.com

8. KIA: The Power to Surprise

Source: http://www.kia.com

9. INFINITI: Inspired Performance

Source: http://www.infiniti.com.tw

10. NISSAN: Innovation That Excites

Source : http://www.nissan.com.tw

11. GUCCI: Quality is remembered long after the price is forgotten

Source : http://www.cucci.com

12. Louis Vuitton: THE TRUTH

Source : http://tw.louisvuitton.com/

13. Hugo Boss: Be your boss

Source: http://www.hugoboss.com

14. Giorgio Armani: It speaks for you.

Source: http://www.armani.com/us

15. Lacoste: Live is a beautiful sport

Source: http://www.lacoste.com

16. DKNY: The official uniform of New York

Source: http://www.dkny.com

17. Prada: Love to come in, don't like to roll.

PRADA

Source: http://www.prada.com

18. H&M: Fashion and quality at the best price

H&M

Source: http://www.hm.com

19. Givenchy: Very irresistible Givenchy

Source: http://www.hm.com

20. Swarovski: Be unlimited

Source: http://www.swarovski.com

Business Slogan (Part 4) 企業標語4

1. Rebook: I am what I am

Source: http://global.reebok.com/

2. New Balance: Let's make excellent happen

Source: http://www.newbalance.com/

3. Converse: Shoes are Boring. Wear Sneakers.

Source: http://www.converse.com

4. **Fila : Power style**

Source: http://www.fila.com

5. **Mizuno: Your passion is our obsession.**

Source: http://www.mizuno.com.tw/

6. **Calvin Klein: Between love and madness lies obsession**

Calvin Klein

Source: http://www.calvinkleininc.com

7. Harley Davidson: American by birth rebel by choice

Source: http://www.harley-davidson.com

8. Marks & Spencer: The customer is always and completely right

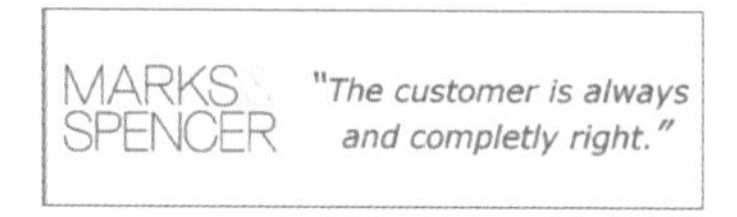

Source: http://www.marksandspencer.com/

9. Johnnie Walker: If you want to impress someone put him on your black list

Source: http://www.johnniewalker.com/

10. Hallmark: When you care enough to send the very best

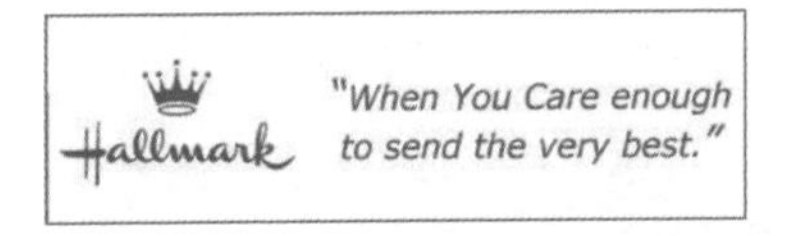

Source: http://www.hallmark.com

11. Tag Heuer: Success. It's a mind game

Source: http://www.tagheuer.com/

12. Maxwell House : Good to the last drop

Source: http://www.maxwellhousecoffee.com

13. AT&T: Reach out and touch someone

Source : http://www.att.com

14. L'Oreal: Because you're worth it

Source : http://www.loreal.com

15. 3M: Innovation

Source : http://www.3m.com

國家圖書館出版品預行編目資料

商用英文寫作／朱海成著.
一 初版. 一 臺北市：五南, 2016.04
面； 公分.
ISBN 978-957-11-8591-0（平裝）

1.商業書信 2.商業英文 3.商業應用文

493.6 105005448

1XOS

商用英文寫作

作　者— 朱海成
發行人— 楊榮川
總編輯— 王翠華
主　編— 朱曉蘋
封面設計— 陳翰陞
出版者— 五南圖書出版股份有限公司
地　址：106台北市大安區和平東路二段339號4樓
電　話：(02)2705-5066　傳　真：(02)2706-6100
網　址：http://www.wunan.com.tw
電子郵件：wunan@wunan.com.tw
劃撥帳號：01068953
戶　名：五南圖書出版股份有限公司
法律顧問　林勝安律師事務所　林勝安律師
出版日期　2016年 4 月初版一刷
定　價　新臺幣450元

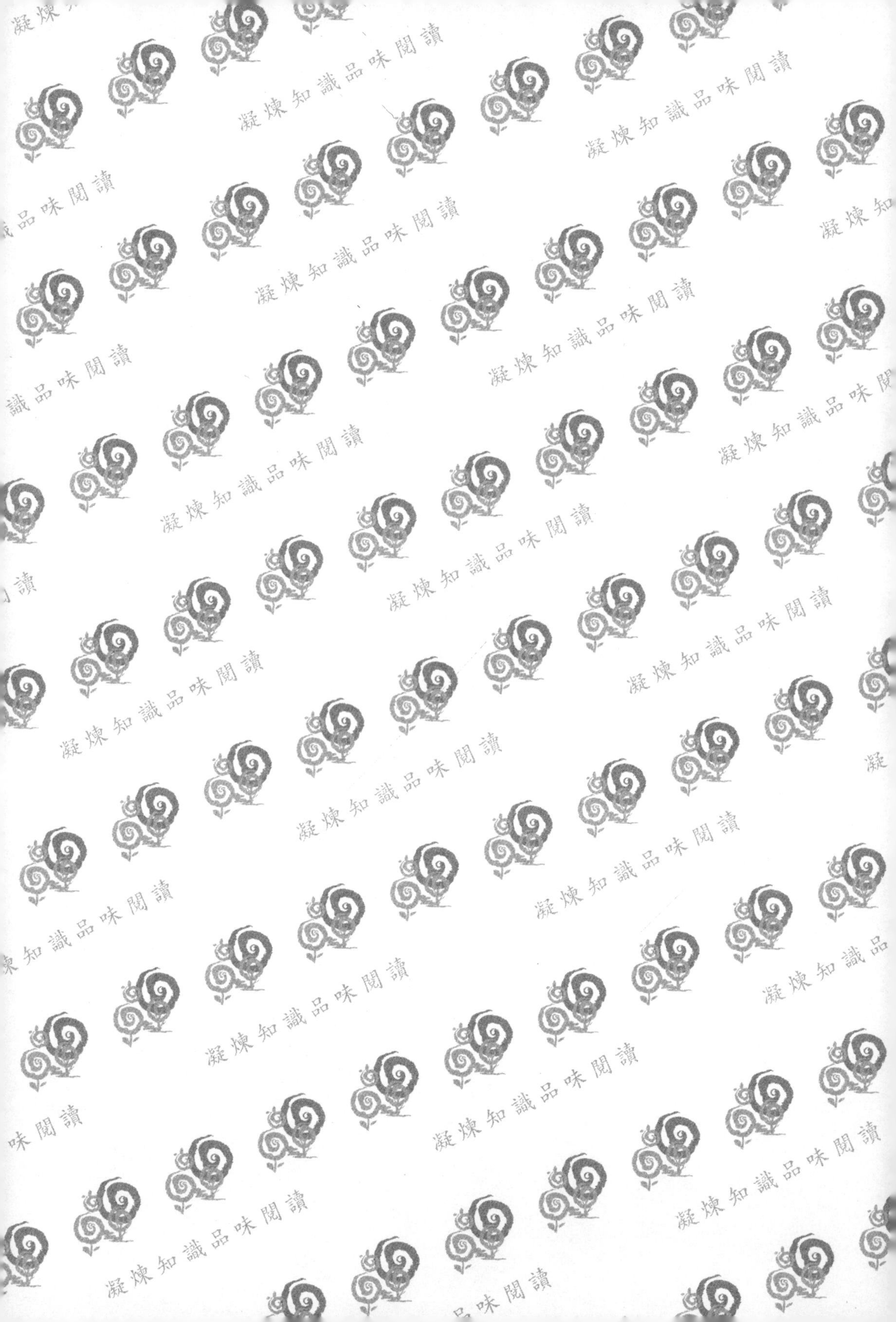

凝煉知識品味閱讀